Scenic Construction
for the Stage

Scenic Construction for the Stage

KEY SKILLS FOR CARPENTERS

Mark Tweed

FOREWORD BY SIR KENNETH BRANAGH

THE CROWOOD PRESS

First published in 2018 by
The Crowood Press Ltd
Ramsbury, Marlborough
Wiltshire SN8 2HR

enquiries@crowood.com

www.crowood.com

This impression 2023

British Library Cataloguing-in-Publication Data
A catalogue record for this book is available from the British Library.

ISBN 978 1 78500 451 3

Frontispiece
Assassins, Jerwood Vanbrugh Theatre, Director: Nona Shepphard, Designer: Judith Croft, Photographer: Linda Carter. © RADA

Typeset and designed by Guy Croton Publishing Services, West Malling, Kent

Printed and bound in India by Thomson Press India Ltd.

CONTENTS

FOREWORD   6

DEDICATION AND ACKNOWLEDGEMENTS   7

INTRODUCTION   9

1   CONSTRUCTION WITHIN A THEATRE PRODUCTION   13

2   THE BUILD PROCESS   27

3   KEY CONSTRUCTION SKILLS   35

4   WORKSHOP MATHS AND GEOMETRY   83

5   FLATS   99

6   TREADS AND STAIRS   131

7   DOORS AND WINDOWS   147

8   STAGE FLOORS, TRUCKS AND PLATFORMS   181

9   ARCHITECTURAL FEATURES   203

GLOSSARY   216

FURTHER READING   221

USEFUL CONTACTS AND SUPPLIERS   222

INDEX   223

INTRODUCTION

ABOUT THIS BOOK

Every carpenter, theatre and workshop will have their own particular style and way of doing things based on any number of factors, so a book that represents the 'definitive' way to build scenery doesn't really exist. However, what I hope this book will provide you with is a solid methodology to enable you to produce scenic elements that are accurate, functional and well finished, whatever the scale or budget. It is designed to encourage a consistent approach, which may be adapted to suit your working practice.

With a consistent approach, it is possible to produce anything that your imagination and resources will allow, and will quickly build confidence to tackle ambitious projects. Whilst the emphasis is very much on practical projects and exercises, there is plenty of background information to put the principles demonstrated within the book into context. As you will see, this approach takes the form of step-by-step projects, technical drawings and sketches, and photographic examples of professional work. Hopefully this will prove a useful resource that will appeal to you whether you are a technical theatre student, professional carpenter, draughtsperson or project manager, and also to those of you who are involved in local theatre groups or school productions, and want to get more involved in creating sets.

Theatre sets are rarely built from a single type of material, and often require input from craftspeople and technicians with a wide variety of skill bases.

In the digital era with projection mapping, automation and VR at the forefront of production design, the very definition of what a 'set' actually comprises changes all the time. However, as advanced and forward looking as theatre production can be, there is always a need for the traditional and the reliable, not only as a foundation from which these new technologies can be used, but to suit the majority of theatre productions, which either require a more traditional aesthetic, or are less reliant on technology.

Carpentry and joinery, as well as metalwork fabrication, represent the mainstay of the industry. In my experience of the two, working in timber is by far the most common practice (although as a carpenter I am obviously biased) as it is an extremely versatile material, and can be undertaken with even the most basic tool kit. That said, timber does require a good understanding of its properties and applications to make the best use of it, which is what I aim to do in this book.

Having a wide range of skills in your repertoire, and the ability to combine different materials, will make you infinitely more adaptable and employable, and I would certainly encourage anyone in this field to try out as many related production crafts as you can. The techniques and materials discussed here should serve as a useful guide for many different projects, not limited purely to scenic construction.

Above all, the most important factor in learning practical skills is developing consistency through practice. No amount of reading will make up for a lack of 'hands-on' practice, so spending time 'on the tools' honing your craft will pay dividends. Put in the time to get the basics right, look for ways to make small improvements, and with patience and

OPPOSITE: An 'A-frame' brace temporarily fitted to a flat in the workshop.

practice you will be producing work of a professional standard.

The aim of this book is to give a solid introduction to the fundamental carpentry skills and processes involved in building scenery for the stage, presented from a practical perspective, with which you can begin to expand your skills and refresh your knowledge.

ABOUT THE AUTHOR

I have always had a passion for making things, and I am never happier than when I am either putting something together or taking it apart. As fun as it can be to make things for yourself, I find that creating something which is part of a collaborative process even more satisfying. Theatre is the perfect arena in which to work with a huge range of people, each with different skills and interests. Together you can produce something which is greater than the sum of its parts, in the sense that you are part of creating an experience that audiences will remember for far longer than just the run of the production.

I trained to be a scenic carpenter at RADA, which gave me an excellent foundation in a wide range of technical skills. I was lucky enough to be offered a job as a production carpenter at the Royal Opera House workshops upon graduation, where I spent a few years honing my skills and broadening my understanding of what went into building for productions on a large scale. My first professional build was on *Die Zauberflöte* (*The Magic Flute*), and I went on to work on many others, including *Faust*, *The Tempest* and *Der Ring des Nibelungen*. The Opera House maintains a traditional approach to scenery in many respects, with carpentry at its core, and during my time there I learned something new every day. I worked with some fantastic people who generously shared

The author, Mark Tweed.

their knowledge and experience with me, much of which I am passing on to you in this book.

Following my time at the Opera House, I returned to RADA as a tutor, and have been the Head of Construction for several years, training students of all abilities, running the department and supervising builds for public productions, of which there are around fifteen per year. Alongside this I continue to work as a freelance carpenter and project manager when time allows! The huge variety of work in the scenic construction field means that every day is different, and every build a chance to work with new people, and either brush up on old skills or try something new.

I hope that reading this book encourages you to push your skills, try new projects and, most importantly, enjoy the work!

Mark

Rookery Nook, Jerwood Vanbrugh Theatre. Director: William Gaskill; Designer: Douglas Heap © RADA.

Set Designer Judith Croft's white card model for RADA's production of *Assassins*.

Designer Adrian Linford and Director Phillip Franks talk through their ideas for *Women Beware Women* with Production Manager Jacqui Leigh.

1
CONSTRUCTION WITHIN A THEATRE PRODUCTION

THE PRODUCTION TEAM

Behind every production is the creative team who provide the aesthetic and artistic setting in which the play can take place. Their process and priorities can be different to that of the workshop, although the team as a whole are working towards the same goal. Understanding the roles within the creative team and being empathetic to their process will ultimately lead to a smoother build and higher standard of work. Making theatre can be a fascinating balance of creative thinking, artistic vision and technical skill, however often the reality of the considerations facing the construction team, such as the budget and schedule, as well as the health and safety considerations and logistics, can at times seem at odds with these aspects.

Developing a deeper appreciation of what can be achieved, and a willingness to negotiate and experiment, rather than just being tempted to do things the 'easiest way' in terms of construction, will aid the finish and your enjoyment. It will mean you create the best work to serve the production, and you will gain greater satisfaction in the work, from being that much more invested in the process as a whole.

The team itself will vary depending on the scale and set-up of a production, but the key roles you may find yourself in contact with are the production manager, the designer and the workshop foreman.

The Production Manager

The production manager is responsible for all the technical aspects of staging a production. They will plan the schedule from the initial design meetings through to the get-out, and need to have a solid understanding of a huge range of both technical and creative disciplines, and the relationship that these have with each other, in order to stage a successful production. In construction terms the production manager will be keeping an eye on the progress of the build in relation to the schedule and budget, as well as keeping the construction department informed of any design changes or developments.

Construction (Design and Management) Regulations 2015, or CDM, is a set of regulations written by the Health and Safety Executive in the UK which now covers most theatres, performance spaces and venues. They specifically focus on putting in place a clear hierarchy of responsibility, from the early stages of the design process through to the management of the fit-up and get-out, relating specifically to sets and staging. The production manager is at the heart of this process. The idea behind it is that systems of communication are streamlined, as all the health and safety information and schedules will be in one place for the whole team to access, and can be updated via a live document.

OPPOSITE: The scenic workshop at the National Theatre, where construction, scenic art and props teams work closely together to produce work of the highest standard.

Jacqui Leigh and Guy Fryer, RADA's Production and Technical Managers respectively, deliver a pre get-out 'toolbox talk' to the technical teams.

How does this affect you as a carpenter? Probably the most important part of this process is the 'toolbox talk' with the team prior to work starting. This will generally be led by the production manager or workshop foreman, and should give everyone a clear picture of the aims for the day. Any details of the plan can be discussed with everyone present, which will make for a smooth and safe build, fit-up or get-out. This should also cover basic requirements, such as where the toilets are, when the breaks are, and who everyone is. This inclusive approach not only keeps everyone informed, but reduces the risk of an accident, saves a lot of time, and avoids the potential for miscommunication.

The Designer

The member of the creative team with whom you should find yourself in regular contact is the designer. In theatre, particularly the designer is often responsible for designing both the set and also costumes to help provide a cohesive style for the piece, although this is not always the case. Their process is closely linked to the play itself, as well as the process of the director.

The designer will begin by reading the play and discussing it with the director, analysing characters and settings to begin to form an idea of the scope of the design. The play may be set in a specific historical period or geographical location, or be staged in a particular style of art or performance. The designer will begin researching these to help provide a creative 'palette' of colours, shapes, objects and locations to inform their design. They will also begin to look at the theatre space itself to get an idea of not only the physical dimensions of the stage, but also the technical possibilities in terms of scene changes, masking, audience configuration and the physical attributes of the venue itself, which may be tied into the design.

Once their research is completed, the designer will begin to produce a 'white card' model box of the set, a 1:25 scale model of the theatre with key elements of set and props modelled simply for experimentation and discussion with the production team. The main purpose of this is to get a sense of the concept in three dimensions, and to see how the design could work in relation to the ground plan. At this point the various practicalities and technical considerations can be discussed with the director, production manager and often the departmental team leaders (construction, scenic art, wardrobe) before committing to a final design.

The designer will produce a final model and supporting drawings to present to the team at a

production meeting. Usually the director will introduce the play and their vision for staging the piece. The designer will present their final design in the form of a finished model with full detail and accurate paint finishes, as well as technical drawings and references for the construction and props teams, and costume drawings and samples for wardrobe to work from.

The play is often 'storyboarded' during the meeting using the model box to show major events and scene changes, and to give the team a good idea of the technical requirements relevant to their department.

So as you can appreciate, the designer's process can be complex, and can encompass a huge range of variables. Therefore supporting them through this process, asking them questions about their design, putting your ideas forward and experimenting with theirs, will lead to interesting and satisfying work.

The Workshop Foreman

As a carpenter, you will generally report directly to the workshop foreman, who will supervise the build taking place in the workshop. They will be a carpentry specialist with an in-depth understanding of the materials, tools and techniques required to undertake the build, and are often the link between the drawing or production team and the fabrication team, making sure that drawings are ready to go to the workshop and that materials are in stock. They will often dictate the pace of the build, keep an eye on the quality of the work before it leaves the workshop, and induct new team members on the use of machinery.

UNDERSTANDING TECHNICAL DRAWINGS

The work of the construction team really begins once the technical drawings and model box are available to work from. These drawings can take many forms, sometimes drawn by hand in a traditional style, and now more commonly using CAD (computer-aided design) programmes. The same applies to the model box, which may be a physical object or may be rendered in 3D to allow for manipulation, again using a CAD programme. It is common for designers to provide references in the form of photographs and sketches to support their drawings, and allow the workshop team to

Hand-drawn designer's plans on the board ready to be analysed by the construction team.

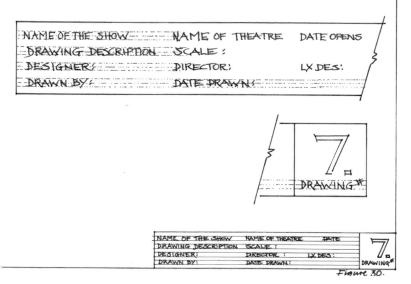

Figure 30.

The title block contains vital information about the technical drawing. GARY THORNE

A scale ruler is an essential tool when interpreting drawings and model boxes.

interpret the style of the piece using the 'language' of the design.

Every designer has their own style, and will express their ideas in a different way depending on the area of their technical expertise. What they all have in common is the desire to communicate their ideas clearly to the team to ensure that their vision of the piece is accurately recreated on stage. Therefore, understanding the format of the most common styles of technical drawing will help you to achieve this goal.

Before reading drawings it is important to understand the visual language and format that they are drawn in. The first port of call when looking at any technical drawing is the title block. This will contain information related to the production such as the name of the play, the director and the venue as well as crucial technical information relating to the drawing itself. Before planning or building anything, be sure to study this carefully, as some of the unanswered questions arising from studying the drawing itself may be addressed here.

In theatre, most printed venue drawings are 1:25 scale (1mm on paper: 25mm in reality); however, some larger theatres, and opera houses in particular, will print at 1:50 due to the size of the

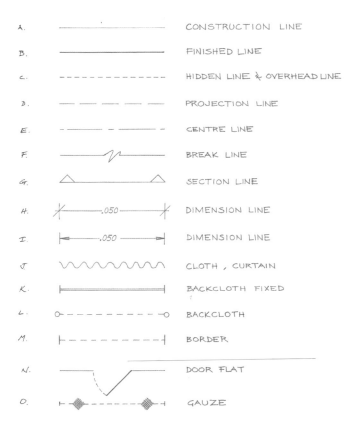

A guide to the various line types found on a technical drawing. GARY THORNE

The use of blood or water effects on stage is good reason to protect the set for the duration of the run. *Phaedra's Love*, GBS Theatre. Director: Iqbal Kahn; designer: Florence de Maré; photographer: Dave Agnew. © RADA

There may also be other factors to protect against, such as water effects, deliberate damage during a fight scene, or perhaps some poor soul appears to bleed to death all over your lovely parquet floor during every performance. Whatever the situation, selecting during the planning stage both the materials in conjunction with the paint finish most appropriate to handle those situations is really important, rather than trying to deal with recurring damage at a later stage. As I mentioned previously, remember that weight or thickness does not equal durability, and the properties of the materials and finish are the main factors here.

Finish

When talking about finish, it is important to understand that this not only concerns the natural finish of the materials you are choosing, but also the implications to the finish that particular methods of fabrication may have. The visible edges of plywood, for example, are hard to disguise due to the layers of alternating grain, so think about how pieces sit in relation to each other, and plan the application of the cladding and visible surfaces carefully, and always in relation to where the audience will be sitting.

Be sympathetic to the design when planning where the breaks in the set will be. With some lateral thinking, joins in the set can be minimized. Look for architectural features to hide joins, such as door frames and moulding, and remember that vertical breaks are generally less noticeable than horizontal ones. Let's look at the following example: for *Hamlet* in RADA's Vanbrugh Theatre, part of the set consisted of a large flown wall, which I broke into four sub-frames for the purposes of access to the stage. Whilst planning the fabrication of this I had to take into account several aspects, including the fact that it was flown, but the main consideration from an aesthetic point of view was minimizing the visual impact of the joins between each section.

As you can see from the image opposite, the wall featured several large windows with substantial moulding. It also featured a sizable cornice with a decorative frieze underneath. Therefore, I decided that the best option was to utilize the window moulding and hide the majority of the break

The main wall for *Hamlet* under construction in RADA's Vanbrugh Theatre. Director: Kenneth Branagh; designer: James Cotterill. © RADA KBTC

lines behind it. These vertical joins would then be masked at the top by the cornice and frieze, leaving only a few hundred millimetres of exposed join, which could be taken care of on stage.

Another consideration to improve the finish can come down to the fixings. For example, cladding a steel frame with thin plywood may require the use of a T-nailer, which, due to the large nail heads, is far less subtle than attaching cladding to a timber frame with a narrow crown stapler. This might not matter, of course – the nail heads could be filled, and there might be texture going on top and no one would be any the wiser – but then again, it could matter a lot. The frame could be part of a minimal set, perhaps even painted white, and the audience could be in close proximity to it. In this case any variations in surface texture from fixings,

sanding and filling would be perceived much more easily by the audience, and a different approach may be required.

It is essential to consider all these factors in discussion with the production team, and not in isolation. There will inevitably be compromise where necessary to achieve a finished set that not only meets the brief of the designer and director, but also meets your budget, and functions well during performances – so remain flexible, and think laterally. Keep several possible ideas 'on the boil' for as long as is necessary, rather than deciding on a single route early on.

Finally, the most important thing to remember above all else is that, no matter how much pressure you are under in terms of time or budget, *never compromise safety*!

2
THE BUILD PROCESS

Working out exactly how elements can be fixed together practically on stage is one of the main areas for consideration when designing and constructing scenic pieces. Building large elements in the comfort of a well lit, well equipped spacious workshop is all well and good, but it is quite rare to be able to simply carry the set through in one piece into the theatre ready for use – unless you are the Royal Opera House, for example, where the entire set can be assembled in the rehearsal room and then simply moved on a huge palette on to the stage! Even then, the set will have been carefully assembled and prefitted in the off-site workshop with transport in mind, so that when it reaches the venue, the crew can simply bolt or screw the elements together knowing that they will fit, with minimal need for cutting or fixing on stage, where time is at a premium due to the demands of the production schedule.

JOINED-UP THINKING

The first step of this process is in the early planning stages. As discussed previously, designing scenery relies on more than just the aesthetic of the finished set and its function during the run of the production: a large proportion of time should be dedicated to thinking about exactly how the set will fit together, and why. As you begin to draw up

the scenic elements, try to imagine the journey of each piece from the workshop to the stage, and try to consider the different requirements to serve the needs of the type of production. A small set for a fringe show, for example, may need to be set up and struck very quickly without anything being fixed down to the stage or walls. A large show at the National Theatre might be in rep, and need to pack up efficiently for storage and then be reassembled consistently each run.

In the case of our productions at RADA, the sets are up for the duration of the run, and our workshops are on site, so the main consideration lies in the logistics of breaking the set into manageable

OPPOSITE: *Ashes and Sand*, Jerwood Vanbrugh Theatre. Director: Jonathan Moore; designer: Neil Irish; photographer: Dave Agnew. © RADA

Sets at the Royal Opera House are packed on to wagons, which can be transported and stored easily.

chunks in order to fit it through the dock door into the GBS theatre, as well as into the goods lift which serves the other two theatre spaces. There-fore not only do we take into account the physical dimensions of each element, but also the weight and the facilities available to move the pieces safely with the team we have available. The same thought process goes for anything you wish to make. Double check the access into each venue or workspace, and also think carefully about how the elements might be transported when planning the build, and incorporate this into the build. Hand holes, eye bolts, travelling battens and castors built into your pieces can make a big difference when it comes to the get-in.

JOINING THE SCENIC ELEMENTS

Once a clear breakdown of the set has been es-tablished, the next step is to decide exactly how the scenic elements will join. There are several reli-able ways of joining scenery on stage that have been tried and tested, and deciding which method best suits your build will be dictated by the exact requirements of the production. There is a massive range of options out there, so the following list is by no means exhaustive but will hopefully get you started. A few favourites are described below.

Backflap Hinges

Backflap hinges are fantastic for joining timber scenery together, as the leaves of the hinge cover a wider area than a butt hinge for strength, and they can be screwed in at different angles on com-plex shapes. It is possible to 'crank' the leaves of the hinge so they form around the edges of timber. For added strength for some applications, such as on a folding gate rostra, they should be attached using machine screws and T-nuts. They are the accepted fixing for much of the work we do, particularly for fixing flats together and attaching

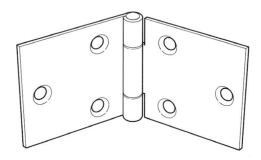

Flints backflap hinge.

braces to scenery. Try to avoid taking screws in and out of the same holes, however, as the wood will quickly deteriorate. If pieces need to separate frequently, then use a pin hinge instead.

Pin Hinge

The pin hinge is basically the same as the backflap hinge, but with a removable pin, the benefit being that timber elements can be easily realigned and fitted without needing to remove screws. Looped

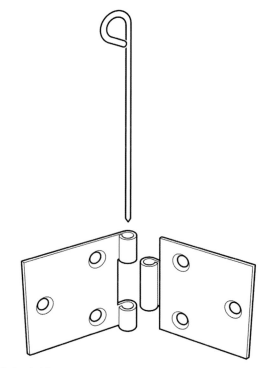

Flints pin hinge.

pins are available to prevent the pins falling out if the hinges are aligned at an awkward angle. They are made with a small amount of play so they can be removed by hand, so are not always the right choice, but they are a really useful and widely used option.

Bolted Lugs

Lugs are a great way of joining pieces as they can easily be bolted or welded on the back of the set with it clamped together, and the joining bolts fitted. If necessary the lugs can be re-aligned easily using a podger (ratchet spanner), and they will withstand a lot of repeated use.

Coach Bolts

Coach bolts are a great option for bolting timber frames and formers together. The coach bolt has a square collar below the domed head, which sinks into the timber preventing it from rotating. This means the nuts can then be tightened from one side. For repeated use it is advisable to use these in conjunction with bolt plates, which will reinforce the timber, acting partly as a crush plate, and also strengthening the square hole for the bolt to fit into.

Split Battens

Split battens consist of a pair of chamfered battens mounted on corresponding edges so they lock together. These are very simple to make, and require no tools when use them, making them ideal for quick turnarounds and scene changes. They are often used on rostra and trucks.

Line and Cleat

Sometimes the simplest traditional approach is the best, particularly when time is against you during a scene change or turnaround. The line and cleat

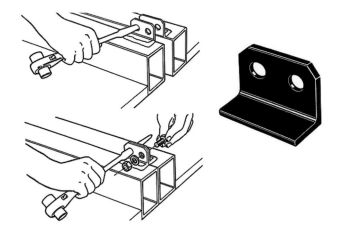

Flints podgalug – a lug for use with the podger.

A coach bolt fitted through a bolt plate.

Split battens make easy work of linking rostra together.

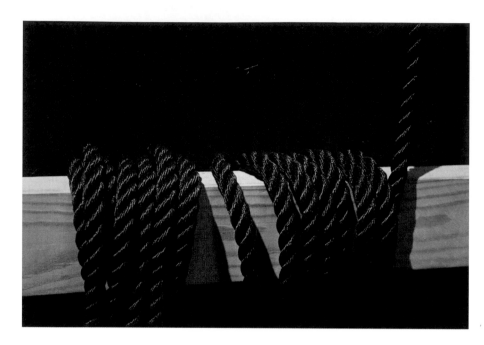

Lines can easily be laced between opposing cleats across flats to join them quickly.

system is a fantastic way of joining tall flats together, essentially by lacing them up on the back with a length of sash. The benefit of this technique is that it can be done from the ground, so there is no need to try and squeeze pins into hinges at the top of a fully extended ladder in the dark – simply throw the line over the cleat at the top, lace it up, and tie it off!

Stud Plates, Peg Plates and Boss Plates

Stud plates, peg plates and boss plates are pieces of hardware that offer a variety of options for bolting scenery together or to the stage. For some situations they should be let into the timber so that the plate sits flush, and then fitted with the appropriate machine screws. This then provides you with a threaded bolt or captive nut for reliable realignment of different parts of the set. There is a huge range of variations of these items for many different applications, so have a good look at the Flints website (flints.co.uk) for more information, as they design and stock the widest range for theatre use.

Labels

The importance of labelling your scenery as you work should not be underestimated! Individual pieces should be labelled on the back with the name of the production, the name of the piece as

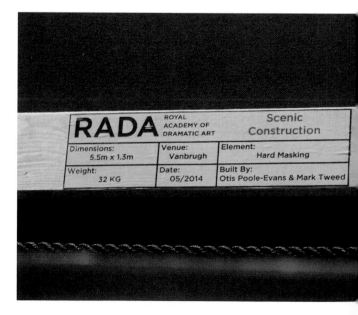

RADA	ROYAL ACADEMY OF DRAMATIC ART	Scenic Construction
Dimensions: 5.5m x 1.3m	Venue: Vanbrugh	Element: Hard Masking
Weight: 32 KG	Date: 05/2014	Built By: Otis Poole-Evans & Mark Tweed

Labels should be clear, accurate and consistent across the set.

Keep it simple! Tie-marking everything will take the guesswork out of reassembly on stage.

it appears on the bench drawing, its orientation, its position on stage, and its weight.

As well as identification labels, using a system of 'tie marks' is recommended to help quickly match up corresponding edges between pieces during fit-up. With two edges lined up, draw a box across both pieces and write the same word on to both edges. Use a different word on each join, and pick a theme for the whole set, such as 'football teams', 'endangered animals', 'types of cheese' or whatever you fancy – though avoid making the theme 'offensive language'!

MOVING THE SET

Pre-Assembly

Check that everything is the correct size once built, as it appears on the construction drawings, cross referenced with the ground plan and section if needs be, prior to it leaving the workshop. When the pressure is on mistakes do happen, and catching them early is important. Any alterations

to the dimensions of flats will be much simpler to remedy in the workshop or assembly area than on stage, and better for scenic art to deal with as well.

Pre-assembling the set prior to the get-in is essential. If you do not have the time or space to construct the entire set in the workshop (as is often the case), then breaking the set down into sub-assemblies and prefitting hardware will save a huge amount of time during the fit-up. It also helps to be able to visualize where any possible areas for snagging may be once the set is constructed, such as unsightly gaps that may need filling, or cutting in, which may need to happen on stage.

The Get-in and Fit-Up

The get-in is the process of moving the set from the workshop to the stage, and the logistics involved in transporting and manoeuvring large pieces of set around safely and efficiently. Once the set is in the space, it can be fitted up by the stage and construction team. This will often include laying a show floor, which may need cutting

A set under construction in the National Theatre assembly area.

in to fit the venue, assembling the main structures and bracing the set, followed by adding details such as hanging doors, fitting moulding, and generally tidying up the set before it is handed over to the production team.

The Get-Out and Safe Deconstruction

There can be a real temptation when the get-out comes round simply to tear up the floor or rip flats apart as quickly as possible; this may be due to being under time pressure, or wanting to be a hero in front of your peers! This is a bad idea for a number of reasons, the main one being that you are putting yourself and your team at risk, as well as making more mess than necessary, which ironically slows the process down. As part of the CDM ('construction design management') process, a clear plan should be established prior to any work starting. This should include a detailed method statement outlining any aspects of the get-out that may be particularly challenging, for example needing to de-rigg heavy pieces of scenery.

A clear timetable and planned order of work should be drawn up for this purpose, usually in conjunction with the production manager, taking into account considerations for other technical departments, and the disposal and storage of the set.

SUSTAINABILITY AND SALVAGING SCENERY

As consumers of natural resources, producers of synthetic materials, and makers of objects that can have limited use but a long lifespan, we need to take responsibility for how we deal with the waste we produce, as well as look at creative ways to reuse the things we have fabricated. There are, of course, many ways to do this – new initiatives are starting all the time, and with the ready accessibility of social media, it is easier than ever to think about how your set can be repurposed, rather than thrown in a skip.

There also needs to be some discussion over the choice of materials used in the first place, as this can have a large influence on how it can be recycled or how durable it may be for reuse.

Materials such as timber and steel are fairly easy to use responsibly, as they can be sourced from sustainable supplies, as well as easily recycled. Plastic, expanding foam and fibreglass, on the other hand, are fantastic materials for set building, but have a lifespan in most cases of thousands of years, so finding cost-effective, ecologically responsible alternatives to some of these is an area the industry needs to explore. There are alternative recycled materials available such as boards made of recycled plastic, although these can be prohibitively expensive; it would take a large increase in usage as well as a few more years of refining manufacturing processes for the cost to be comparable to that of timber and steel.

Theatre, and indeed the entertainment industry as a whole, is big business, and sadly it can be 'better business', particularly on larger scale work, to scrap sets and build from scratch rather than to transport, store and reuse. That mentality is beginning to shift, however, and many leaders around the theatre industry are looking at initiatives to make sustainability more of a priority.

One of the fantastic aspects of theatre is being able to engage with it from a grass roots amateur level, maybe a local theatre or school production, all the way up to big West End and Broadway shows, opera and live events. There is already a huge network within this range of people, creating new work of every style and budget, and so with a bit of planning, creativity and networking, it is entirely possible to source very specific items of set as well as find someone who may be able to make use of pieces you are considering putting in the skip. This is perhaps the way in which it is possible to make the biggest impact in terms of sustainability from a personal perspective: all it takes is bit of effort.

In practical terms there are several ways in which to make your build eco-friendlier and help out others:

- Where practical, consider building the set in recycled materials or repurposing used set elements rather than building from scratch. For tips on how to modify an existing flat, see the flats chapter in this book
- Advertise on social media any items of set that are intended for the skip: many theatres, schools and societies would be glad of anything they can get their hands on
- Before scrapping anything, strip any reusable materials and hardware from items of set, cut out rails, saw out ply panels, take off moulding and add it to your workshop stock
- Ensure that the materials you intend to use are from sustainable supplies; in the case of timber, FSC certification ensures that there was no negative impact to the forests from where the material was sourced. Also consider the carbon footprint of transporting the material to your workshop
- Refer to the sustainability section at the back of this book for some useful contacts

3
KEY CONSTRUCTION SKILLS

HEALTH AND SAFETY

Above all else, your personal safety is the most important consideration in the workshop. Make no mistake, building scenery – and indeed, working in the theatre in general – can be a hazardous profession at times. There are many systems put in place by employers to reduce the potential risks and promote safe working practice, but taking your own safety seriously is the only way to ensure this. Investing in, and getting into the habit of wearing PPE (personal protective equipment) will ensure that you enjoy a full, injury-free career. Some employers will provide PPE depending on the set-up, but it is advisable to invest in your own so that there is no situation where you are not suitably equipped for working safely. You are also more likely to wear your PPE if it is comfortable and fits well – second-hand dust masks, for example, are not very appealing!

There are many different options, but as a minimum you should have the following:

Safety specs/goggles: Your eyes are extremely vulnerable to splinters, dust and sparks, so protect them!

Safety footwear: Moving heavy materials and equipment can cause broken toes and twisted ankles, so investing in a decent pair of boots or shoes

OPPOSITE: Clayton Handley, an experienced carpenter at the National Theatre, working on the bench.

with reinforced toecaps will protect them from harm. Remember you will be wearing them all day, so choose wisely! Personally I would avoid rigging boots unless you are spending the majority of your working day outside, as they can get very hot.

Ear defenders/plugs: Working in a typical workshop environment, where the sound levels regularly reach over 85dB, can quickly affect your hearing. Wear moulded plugs for all day comfort, or headphone-type defenders for shorter stints.

Dust mask: Inhaling any type of particulate, be it wood dust, paint, plastic or welding fumes, will take its toll on your lungs, so choosing a suitable mask is important. For example, I would recommend using a P2-rated mask as a minimum for decent protection whilst sanding or using a router. Most modern workshop equipment will be fitted with a form of LEV (local exhaust ventilation), which will extract much of the dust, but when working on the bench or on stage where dust control is more difficult, then wearing a mask is crucial.

Gloves: I would strongly advise against wearing gloves when using machinery or hand tools due to the risk of them getting caught in moving parts; however, they are essential when moving heavy materials and pieces of scenery around your workspace, or using products that fall under COSHH ('Control of Substances Hazardous to Health') such as adhesives and expanding foam. Therefore have a pair of rigging gloves and a box of latex gloves in your kit as a minimum.

PPE is an essential part of working in construction.

HAVING THE RIGHT TOOLS

It might sound obvious, but having the right tools to hand when you need them makes a world of difference to your productivity and accuracy. It is not about going out and buying all of the tools on the list, or necessarily buying the 'best' or most expensive tools either, but thinking carefully about what you are likely to need from day one, looking at what is already available to you, and building your tool kit gradually as you learn what works for you, and what tools you require for the kind of work you are doing.

Looking after your tools is important: the workshop and stage environment can quickly take its toll on your kit if you don't look after it, so investing in a decent toolbox or bag is an essential start. If you are lucky enough to work on your own bench, then fitting out a drawer with dividers for tools is time well spent, as it will then be easy to keep everything organized and to find it when needed.

Label *everything*. Working in theatre is a collaborative effort – people will borrow your things and you will borrow theirs, and it will be difficult to track down any of your tools if they have no distinguishing features. It is common practice either to choose a unique colour combination of electrical tape to wrap your tools with, or to engrave them with your name, or to spray paint them to make them as unique as possible.

Keep your tools sharp. Nothing compromises the finish of your work more than working with tools that are blunt: not only will they cause damage and wayward cuts, but they could cause an accident if they slip or break. Quality can be a factor, but is by no means the only consideration; establishing a routine of sharpening your tools once a week or whenever there is 'down time' will make the biggest difference to your work. Look at it this way: a sharp chisel that cost £5 is going to be a lot more useful than a blunt £50 one!

Plane irons, chisels and knives should be regularly honed to keep a keen edge; also wiping dust and glue off them before putting them away will prolong their life considerably. The dust will absorb moisture from the air over time and cause corrosion, and glue is obviously much easier to clean off before it has cured. Get into the habit of putting your chisels back in their roll, and store your planes either on their side or with the sole raised, as this will help to protect their edges.

In case you weren't sure, that is definitely my nail gun.

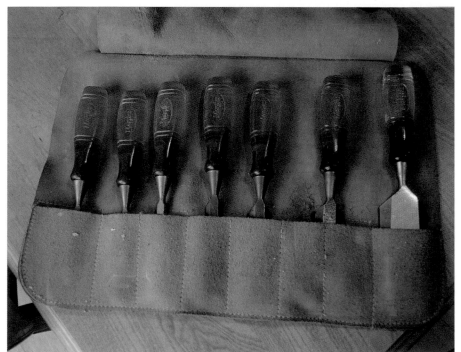

Get rid of the end caps from your chisels and keep them protected in a roll.

- The higher the voltage (V), the more torque the drill will have
- Look out for a 13mm chuck capacity to suit the full range of drill bits you will likely use on a daily basis

You will also find the tools in the following pictures are used throughout the book, and allow most of the projects to be completed without workshop machinery:

A router with a ½in collet like this can cope with most tasks, but a ¼in laminate trimmer is useful for more intricate work.

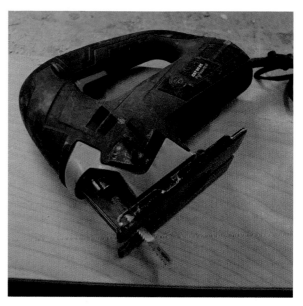

Be sure to change the blade on the jigsaw regularly for neat, consistent cuts.

The plunge saw is one of the best types available, and definitely worth the investment.

While the belt sander is good for removing large amounts of material quickly, the orbital is the one to choose to obtain a smooth finish prior to painting.

Machines

You will need to be proficient in the use of many different workshop machines, so be sure to get a full induction on these when starting out in a new workshop. Although similar in operation, the exact set-up will vary a lot between manufacturers, so do a couple of test cuts on each machine until you are comfortable. Whenever using a machine be sure it is clear of waste, and reset to its default position afterwards. Have a push stick nearby if you want to keep a full complement of fingers!

The main machines you will use on a regular basis are shown in the following pictures; you will need them to complete some of the projects in this book.

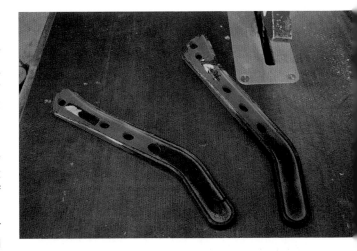

A push stick's only function is to protect your fingers from certain damage, so always have one nearby.

Cross cut saw: Allows for accurate 90-degree cuts across the grain, ideal for cutting timber and narrower rips of plywood and MDF to length.

Table saw: Perfect for ripping timber and boards along the grain. The blade height and angle can be adjusted to suit different materials and tasks.

Bandsaw: The relatively fine teeth and flexible blade allow for more organic shapes to be cut with ease. The vertical movement of the blade means smaller parts can be worked on safely.

Vertical panel saw or wall saw: Very useful for dimensioning large sheet materials which are fully supported on the frame, it can be set to make accurate vertical or horizontal cuts by moving the saw head, making it more user friendly than the table saw for larger work.

Compound mitre saw: Used for cross cutting angles into timber, as well as a very useful tool to use as a compact cross- cut saw when working outside of the workshop or where space is limited. The blade can usually be set to any angle up to 45 degrees in both the X and Y axis.

Tenoner: For workshops regularly producing mortise and tenon flats, this is an essential machine. The tenon length and thickness can be adjusted to suit a wide range of timber sizes.

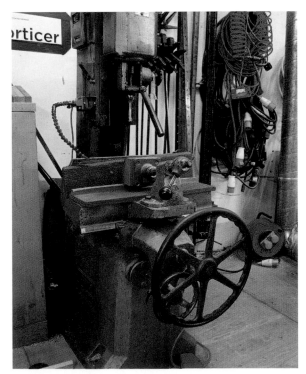

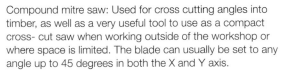

Mortiser: Easily the most efficient way of machining mortises, the depth of cut as well as the horizontal position of the chisel can be adjusted. A variety of chisels are available to suit most timber and tenon sizes.

Planer thicknesser: The overhand planer can be used to square up rough timber, and the thicknesser is perfect for dressing timber stock to a specific size with consistent accuracy.

Many of the functions of these can also be performed using hand and power tools, so if you have limited access to machinery then don't be put off, as there are many ways to get the job done! Over-reliance on machinery is a common mistake for beginners, and it is often better to learn to build with hand and power tools first, as it develops accuracy.

SHARPENING

There are many different methods for sharpening and honing chisels and planes, and there is much debate over which is superior. Ultimately a sharp edge is what you need, so for practicality we will look at a simple, reliable method that will be suitable for the majority of tasks required of a stage carpenter. We are generally not dealing with exotic hardwoods, nor fine cabinetry levels of detail, so spending days achieving a mirror finish on your tools, whilst a great skill, is probably a little 'over the top' when a reliable sharp edge will suffice. Like learning any skill, patience and practice will make the process more efficient and accurate, so sharpening 'little and often' is the best way to start out.

There are three main stages to sharpening: grinding, honing and lapping.

Grinding

When an edge is visibly damaged with nicks and chips, or it is visibly out of square, it will need grinding back to a clean square edge before honing. This can be achieved on a diamond or oil stone; however, a grinding wheel is generally preferable for this, as it takes seconds rather than several minutes. Care is needed on the wheel to ensure that the blade does not overheat, which will prevent the edge from staying sharp.

Honing

Next the bevel is shaped either flat or convex, depending on which style of edge you choose. There are several styles, but the two most common are 'micro bevel' and 'convex bevel'.

Keeping your hand tools sharp will make a big difference to your work.

This chisel is more useful as a teaspoon than a sharp edge.

A micro bevel of 30 degrees is ground on to the main bevel of 25 degrees.

The convex bevel is a more freehand approach, but equally as sharp.

'MICRO BEVEL'

The flat bevel will have a 'micro bevel', or a second, slightly steeper angle honed on to the edge, so must be done in two stages. The benefit of this method is that resharpening is generally just a case of reworking the micro bevel rather than having to hone the primary bevel totally flat every time. The down side is that micro bevels can be more easily damaged and chipped as they can be very thin. Honing guides are perfect for this style and will help a great deal when starting out, as maintaining a consistent angle by hand can be very challenging. Many honing guides state how far the iron or chisel should project out in order to maintain the correct angle to make consistent sharpening easier.

'CONVEX BEVEL'

A simpler, more freehand method for sharpening involves shaping the bevel into a slightly convex curve. The idea behind this is to negate the need for a micro bevel, as the edge will have more support as well as making sharpening without a guide much easier. After exclusively using the micro-bevel method for many years, I have been increasingly using this technique as I find it much more efficient when dealing with a whole set of chisels and planes. I also find working without a honing guide to be more comfortable, as it allows for a more natural movement, as well as not having to worry so much about keeping specific angles – but it does take a bit of practice to get consistent. My first few attempts were far from perfect, to be honest!

Either of these styles will give you reliable sharp edges, but the focus of this book is key skills, so I have focused on the micro-bevel method due to the fact that most new chisels and planes are ground with this in mind, and using a honing guide to start out will give more reliable results.

Lapping

Very much like cutting metal, honing produces an unwanted burr or waste material on the cutting edge, which must be dealt with without blunting

Lapping on a leather strop.

the honed bevel and having to start again! This is where lapping comes in. Using either the stone or a leather strop produces a sharp and clean finished edge, and is the crucial final step for really sharp tools!

Sharpening a Chisel or Plane Iron

You will need the following:

- Bevel-edged chisel (any width, but 25mm is a good size to start)
- Plane iron
- Honing guide
- Diamond stone with both coarse (400 grit) and fine (1,000 grit) sides
- Small spray bottle of water
- Leather strop
- Honing compound
- Paper towel or rag

The steps below apply to both the chisel and the plane iron, the only difference being the position of each in the honing guide, as shown below.

Note the position of the chisel in the centre of the honing guide, compared with the plane iron in the top.

This tutorial features a brand new, inexpensive 25mm bevelled-edge chisel. For resharpening used chisels you may want to skip Step 1.

A new chisel ready to be sharpened.

STEP 1

The chisel is brand new and factory ground, so as you can see in the photo, there is no sharp edge, no micro bevel, and very obvious striations where it has been machined on both the front and back of the chisel. The first stage is to prepare the chisel by flattening the back.

Ensure that your diamond stone is firmly secured to the bench or in a vice, coarse side up, and spray the stone with water. Hold the back of the chisel with both hands flat on the stone and rotated slightly, and work it back and forth ten to fifteen times. Turn the chisel over and wipe it off, then you can begin to see where the stone has started to polish the back of the chisel. It is quite common for there to be a slight hollow in the back of the chisel: this is nothing to worry about; the important thing is that the first 15–20mm from the end of the chisel is consistently flat, so keep going until you have achieved this.

Now flip the diamond stone over to the fine side and wet it, and repeat the process, this time working the chisel for longer, perhaps twenty to thirty times, and rotated at a slightly different angle, but always flat on the stone. When you are happy with

Step 1a: Initially flattening the back.

Step 1b: The end and edges are flat enough for sharpening.

Step 2a: The chisel is mounted bevel side down in the honing guide.

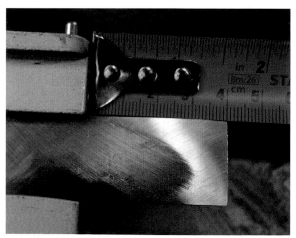

Step 2b: The projection is set at 40mm to grind the primary bevel correctly.

the result, wipe the chisel off with your cloth. To polish it further, rub some honing compound on to a scrap of timber and buff the back, then wipe the chisel off again.

STEP 2

Now that the chisel is prepared, the next stage is to work on the bevel. We are going to use a honing guide, which will maintain a consistent angle throughout. Flip the chisel over so that the bevel side is facing down, and feed it through the honing guide, then gently hand tighten it in place.

The important factor here is setting the projection of the chisel from the guide, which is the amount that the chisel sticks out from the edge. This is important, as it will dictate the angle at which the bevel is honed. The easiest way to set this is to try to match the bevel that is already ground on the chisel. This will require the least amount of work to clean up. Some honing guides have a suggested projection distance printed on them, so you may wish to do this instead.

Regarding angles, the primary bevel should be around 25 degrees and the secondary bevel around 30 degrees, but this does not need to be exactly followed, it is just a good guide for general use. As an example, with my honing guide the primary bevel on a chisel requires a projection of 38mm (for 25 degrees) and the secondary bevel 30mm (for 30 degrees). The projection will be different for a plane iron which is held in the upper part of the guide, though the angles are the same.

To easily set the projection, place the guide with the chisel in it on a flat surface, ideally the bed of a table saw for example, and fine tune its position until the factory bevel is sitting flat. Tighten the guide up with a flat blade screwdriver.

Wet the coarse side of the stone, and begin working the chisel and guide back and forth applying firm downward pressure. Every fifteen to twenty passes, check the face of the bevel, and repeat until the entire face is evenly ground and there are no signs of the factory marks. By this point a burr of metal should have started to develop on the back edge. If this is the same width as the chisel it is a good sign that the edge is evenly ground. We will remove this later.

Now flip the stone over to the fine side and repeat the process, increasing the number of passes until the marks from the coarse stone have disappeared.

Step 3

Now for the micro bevel. Pull the chisel back through the honing guide by about 10mm and retighten it. This will steepen the angle for the cutting edge. This time we don't need the coarse side of the stone, we will just use the fine side, so apply a splash of water and work the chisel back and forth once more. This time we only need to create an edge which is 2-3mm deep at the steeper angle, so don't get too carried away. Do 5-10 passes at a time, wipe the chisel off and check the edge, and repeat until it is even.

Now remove the chisel from the honing guide: we want to remove the burr carefully from the edge. Hold the back of the chisel absolutely flat on the stone and rotated at around 45 degrees in relation to the side, applying firm pressure with both hands. Pull the chisel back, away from the sharp edge, which will remove the burr. Don't be tempted to push it back and forth as this will blunt the edge.

Step 4

The last stage is lapping, using the leather strop and honing compound to finish the edge. This will polish it and remove any last traces of the burr. Clamp the leather strop into the vice, and 'charge' it with the compound. Hold the chisel bevel side down on the leather, and pull it backwards several times. Repeat this with the back of the chisel flat on the leather, and then again for the last time on the bevel side. Wipe off any excess muck and honing compound, and bask in the glory of a fantastically sharp chisel (or plane iron).

Try to keep the edges of your chisels and planes sharp by freshening up the edge regularly. There is no need to regrind the entire bevel each time, simply repeat Steps 3 and 4, unless there are visible nicks in the edge, in which case a clean edge will need to be obtained again by regrinding.

Step 2c: The primary bevel has been evenly ground.

Step 3a: The secondary bevel is ground…

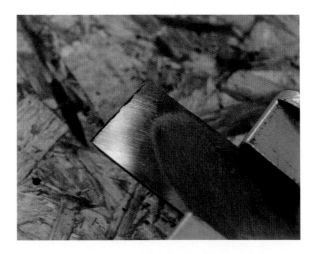

Step 3b: …but the burr must be removed prior to use.

Step 4.

Your sharpened chisel should cut into the timber with ease, but mind those fingers!

Jigs

Jigs are essentially bespoke accessories designed and made to achieve specific workshop tasks using existing equipment. The two jigs I use the most are the peg jig for the bandsaw, and the toggle-shoe jig for the table saw, but these are just two examples. In reality you can make a jig for nearly any construction process, with a bit of planning: it really comes down to which steps of the process you are trying to streamline. Ultimately if it takes longer to set up the jig than simply perform the task, then it is probably a waste of time, but if it allows you to push the capabilities of a machine or to mass produce components easily, then it is definitely worth spending the time on. Below are just two examples of jigs I use to make mortise-and-tenon flats, but there are hundreds of possibilities.

How to Make Toggle Shoes

The benefit of using toggle shoes on taller mortise-and-tenon flats is that they allow rails to be screwed into the stiles rather than mortised directly into the stile, which may weaken it over a long length. 'Blind' toggle shoes, without a mortise, are also useful in some situations, such as adding support under profiled cladding, or as feet on a timber rostra frame, so I generally have a box of them in the workshop.

The jig is simply a plywood board with a slot cut into it at 22 degrees, wide enough to fit a length of 3 × 1in timber into it, fitted with handles so it can be run against the fence on the table saw.

Step 1: Set the table-saw fence to suit the jig, and adjust the blade height for 3 × 1in PAR. Cut a length of timber at 450mm, and slot it into the jig.

Step 2: Carefully run the timber through the saw, cutting off the shoulder of the piece, then flip the timber over and repeat so both shoulders are cut evenly.

Every workshop should have a collection of bespoke jigs.

Step 1.

Step 3: Mark out a mortise width across the shoes, machine them and clean them out.

Step 4: The shoes can now be drilled and countersunk ready for fitting.

The peg jig is simply a plywood plate with an angled notch to fit a 100mm-long timber block into it, to enable you to rip the block into angled wedges on the bandsaw. The wedges are then run through, the wide end at the back creating pegs for joining mortise-and-tenon flats. The offcuts make useful glue spatulas, too!

Step 2.

Step 3.

Step 4.

Courtney cutting pegs on the bandsaw using a bespoke jig.

GENERAL KEY SKILLS

Accuracy

Accuracy when measuring and cutting is a fundamental skill that will have the biggest effect on the quality of your work over anything else, and can easily be taken for granted. Like any practical skill, practice is crucial, as is criticising your work as you go and making corrections if you can see something has been cut inaccurately. No matter how neat your finish is, or how methodical your process, or how quickly you are able to put something together, if your cutting isn't millimetre perfect, then more practice is needed. Sets, of course, will involve multiple flats and assemblies, and rely on tight tolerances in order to function, so whilst a couple of millimetres here and there may not seem too important on an individual piece, when that inaccuracy is multiplied across several pieces, it can cause unnecessary problems during the fit-up. I have been guilty myself of saying a few times 'that'll be all right' if something is a few millimetres out, and then paying the price further down the line when the flats don't fit together; this is an easy trap to fall into, particularly when working to a tight deadline.

As a general rule, avoid drawing around other pieces or using them to mark other lengths, as the marks will inevitably be slightly 'off' the original. Using an end stop on a crosscut saw, for example, is a great way of cutting multiples of the same length accurately.

Accuracy when cutting can be improved by ensuring that any marks indicating the length are clearly marked using a sharp pencil, and also that lines are accurately squared across the face of the material to avoid wayward cuts. Understanding the tools you are using will also help. Saw blades, for example, have a 'set' on them, which means that the teeth splay out on either side of the body of the blade. Therefore, aligning your marks to the very widest part of the tooth on the blade will ensure that the cut is spot on. Blade thickness is a factor in the sense that it is important to make sure that the cut is into the 'offcut' side of the piece,

End stops ensure accuracy when cutting multiple pieces of the same length.

METRIC VERSUS IMPERIAL

Despite the fact that we work in metric here in the UK, many of the stock sizes for timber and boards are still referred to in imperial terms, mainly as a way of identifying them easily (metric dimensions don't make for very catchy names!). You will notice that I refer to sheet materials as '8 × 4s', or timber as '3 × 1in', for example, throughout the book.

Most timber used for scenic construction is referred to as 'PAR', which stands for 'planed all round', which means that if you actually measure a piece of 3 × 1in PAR, for example, instead of measuring 76.2 × 25.4mm, which would be the direct conversion, its dressed size is usually around 70 × 22mm, as it has been planed down. If in doubt, double check the actual metric dimensions of the finished timber before calculating cutting lists; this will vary slightly between suppliers, and even between batches.

Plywood is the best sheet material for structural applications.

about which face of the timber will be seen, and how this might affect the look or function of the finished piece.

Always face off your timber before you start to use a length, by cutting off a few millimetres from the end. This ensures it is square, and gives you a nicely even edge to work from.

There are several sizes of timber that we stock as standard in the workshop. These are as follows:

3 × 1in PAR (70 × 22mm): Ideal for flats, and the main size that we use across a wide range of tasks.

2 × 1in (44 × 22mm): Useful for smaller frames and corner braces.

3 × 1¼in PAR (70 × 28mm): Good for tall mortise-and-tenon flats, French braces, and handrails.

6 × 1¼in PAR (145 × 28mm): Used for heavier duty applications such truck bases and platforms.

One of the fantastic properties of timber is that it is very easy to machine in order to create whatever size stock you may need for a job, so the list above is a good starting point to enable you to build the projects in this book and tackle many different set builds.

PLYWOOD

Plywood is an extremely versatile sheet material for scenic use, and in conjunction with timber, forms the mainstay of much of what we build. It is available in a wide range of thicknesses and grades, and in a variety of useful sizes. Making the choice of which variant to use ultimately comes down to the type of wood used to manufacture the plywood, how it will be used, and the price.

Plywood is made up of multiple layers known as plys, which are of varying thickness and number depending on the type of plywood. Each ply layer is laid with the grain running at 90 degrees to the next, which is what gives plywood its reliable strength and stability. A sheet can be cut in any direction and retain most of its strength. It is also very strong when placed on edge, making it an ideal material for making many three-dimensional shapes. The down side of plywood can be the finish. Due to the layered look of the edges of the board, it can be difficult to achieve an acceptable finish, and the edges won't take paint particularly well, so generally they will be kept hidden from the audience. The grain direction on the faces of the sheet may also need consideration in terms of finish, but many types of ply are available in both long or short grain.

18mm (¾in approximately) thickness is perfect for many structural applications such as treads and rostra tops. I generally use the Finnish spruce Wisa, which is a softwood version and a good balance of strength, weight and affordability, or birch if I am focusing on a cleaner, knot-free finish or improved durability. A good intermediate choice is 'Far Eastern' ply, which strikes a balance between the two, but if being more eco-friendly is a priority, then a good start is to source your boards from nearer the UK (or wherever you might be!).

Plywood can be a great material to clad flats with, as it is lighter and more durable compared with MDF. For this application, thinner sizes such as 4mm or 6mm are ideal, and hardwood faced to obtain a smooth, knot-free finish.

Flexible plywood is a really useful material for curved work; I generally use it in thicknesses of 5mm or 12mm, depending on the application, and it can be laminated together for added strength.

Skin ply is around 1.5mm thick, generally birch based, and a nice material for smaller curves and intricate work. It comes in smaller sheet sizes than most, of around 5 × 5ft, but is quite delicate compared with flexi-ply.

Complex shapes can be constructed by using plywood formers, like these curved rostra.

MDF

MDF, or medium-density fibreboard, is essentially a sheet material made up of wood fibres mixed with a resin-based adhesive, and formed under high temperature and pressure. This results in a board that is very smooth on both faces, with a slightly porous-looking edge. It is very useful for finished surfaces, and gets widely used for cladding as well as for fabricating non-structural 3D objects where a smooth finish, rather than durability, is a priority. A wide range of moulding is also available in primed MDF lengths, which are great as there are no knots to work around. It is easy to shape and cut, and has no directional grain to worry about – although screwing into the edge should be avoided without careful pilot holes. The other positive factor about MDF is that it is very inexpensive in relation to plywood, for example.

However, in my opinion there are a few drawbacks to using MDF. The first and most important is that it is not very good for you. To put that in some sort of context, cutting anything that

MDF is a useful material for certain applications, but is not very durable.

produces dust has the potential to be harmful to your health if not properly contained, but due to the way in which MDF is manufactured, the dust from it is extremely fine. This means it stays airborne for longer, and also has the potential to cause skin irritation and respiratory problems. So before using it, be sure that you have sufficient LEV kit and PPE so that working with it is manageable on a day-to-day basis.

The other negative in my view is its limited durability. Edges and corners are quite vulnerable to damage, and can easily split and tear, therefore careful handling of finished pieces is required, particularly given its weight.

Plastic

For scenery the main type of plastic to use is solid polycarbonate due to its high strength and fire-retardant properties. It is available in a wide range of colours, opacities and thicknesses for many different applications, including floors and glazing for windows.

In recent years twin and multiwall polycarbonate has become a popular choice for cladding some scenery, due to the fact that it is extremely light and rigid, and comes in large sheets, often suitable for cladding whole flats in a single piece. It has some drawbacks in that the edges need covering if visible due to the structure of the sheet, and some types have a ribbed surface that doesn't lend itself to a flat finish. It can be screwed to frames by way of a countersunk pilot hole, but care must be taken that the screw doesn't pull all the way through the sheet.

Vacuum-formed plastic is used for adding intricate replica mouldings, carvings and textures to sets; it is generally either 0.5mm PVC, or thicker 1.5mm fire-retardant ABS for heavier duty work. Peter Evans Studios have a fantastic catalogue of all kinds of different shapes, and are definitely the first port of call for anything vacuum formed.

Polycarbonate is a good choice of plastic for stage use.

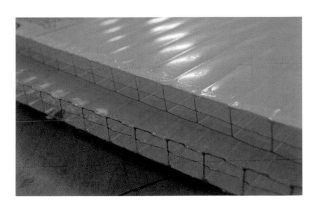

Multiwall polycarbonate is very light in weight and comes in large sheets, making it a good choice for some large set pieces.

Vac-formed plastic, such as these brick-patterned sheets, is a cost-effective and reliable way of adding texture and depth to sets.

Sheet Joins

One of the most common considerations you will need to make as a scenic carpenter is how to deal with sheet joins properly. Whether cladding a flat using multiple boards or extending rips of ply to make a reveal, knowing how to support the join in each case will ensure a sturdy build and neat finish.

Cladding seams in flats are generally dealt with in the planning stage by adding in support rails where the sheets will meet. These will be set face on to the flat to allow a wider surface area to attach the staples. With a flat built 'on edge' it is necessary to notch in rails for this purpose. Essentially this is still a lap joint, but it can easily be cut using a jigsaw prior to assembly.

Due to the standard sizes of boards such as plywood and MDF, you will often need to plate these joints to extend lengths for support. Where it is not practical to add in a support rail, a plate cut to sit generously over both sides of the joint is required. This should be cut back from any edges by 20mm or so (or set back further where hardware needs to fit), and drilled and countersunk with screws every 70mm or so.

The joint can be pulled tight prior to plating using corrugated dogs or 'wiggle pins' as they are also known, but these fixings will not be sufficient on their own.

The plate can then be fixed using glue and screws. I like to use 5 x 30mm screws for plating 18mm material as they give substantial support.

Notching in rails is essential for adding support to flats for cladding seams.

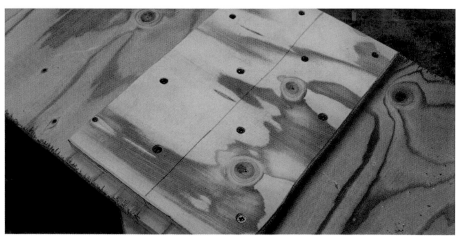

Spacing screws evenly is best practice. Don't let anyone tell you otherwise.

Wiggle pins help to pull the join together, but always need a plate to give the joint strength.

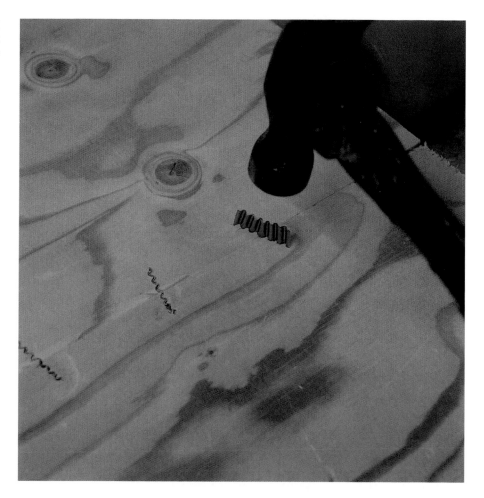

FLYING HARDWARE

Flats that will be flown or suspended should be fitted with appropriate hardware prior to leaving the workshop. Typically, a flown flat will have flying irons at the bottom, with a series of grommets fitted to the stiles and top rail to guide steel wire ropes through. The exact specification and full range of fittings can be found on the Flints website or in their catalogue (everyone should own one!), but any hardware fitted to scenery or set joined together which is suspended above the stage or auditorium should be attached using mechanical fixings, usually bolts – screws are not enough! If you are not sure as to whether something is safe, always seek professional advice, and don't leave anything to chance. In the example below I am fitting an opera house grummet to a 3 × 1in stile. This method can be applied to many different types of Flints fittings, but be sure to check the specification for bolt sizes and working load limits (WLL) before starting, and always consult an expert before rigging or flying anything.

You will need the following:

- Flints grummet
- Combination square
- 5mm drill bit
- 6.5mm drill bit
- M5 countersunk posi-drive machine screws
- M5 T-nuts
- PZ2 screwdriver
- Hammer

Method

Step 1

Position the back plate of the grummet so it is sitting perfectly square on the stile, and mark the bolt positions with a pencil. Drill out the four holes using a 5mm drill bit: be sure to put a piece of scrap underneath to prevent breakout.

Step 2

On the opposite side of the stile, open out the bolt holes a little using the 6.5mm drill, but don't drill all the way through. We just want to make space

Step 1.

Flints sell an excellent range of hardware specifically for use on stage. This Opera House grummet is designed to allow the ferrules on made-up wire ropes to pass through the slot for quick turnarounds.

Step 2.

Step 3.

for the threaded collar of the T-nut, so half depth should be plenty. Position a T-nut over each hole, making sure there is support underneath from the bench or trestles, and hammer them as flush as possible into the surface of the stile.

STEP 3

Now fit the grummet to the opposite side, with a machine screw in each T-nut. In my experience, using an impact driver to tighten these is not always the best choice, as the heads on some machine screws can easily be damaged, at which point they become difficult to remove or tighten any further. I use a regular short-handled PZ2 screwdriver instead, with which it is possible to exert more pressure and prevent slipping to pull the grummet tight to the stile. If necessary, cut any protruding threads flush on the back, and the grummet is ready for use.

4
WORKSHOP MATHS
AND GEOMETRY

After finishing at school I wondered how often I would need to know how to calculate the circumference of a circle, or to use Pythagoras' theorem to find the missing length in a triangle – but as it turns out, it is indeed quite often!

Whilst using AutoCAD can go a long way to help calculate lengths and angles, and to unlock complex shapes, there is still a need to know and practise basic maths and geometry whilst working on the bench, both as a means to mark out, as well as to double check work and calculate material usage. The following section is really just to recap some of those formulae and methods you may well already know, and to put them in a practical context to use day to day in the workshop. These are the very basics, a quick reference guide that may be useful when tackling some of the projects in this book, for example.

CIRCLES

The following are some circle basics:

- The distance across a circle through the centre is known as the diameter, or d
- The distance across a circle from the edge to the centre is known as the radius, or r
- The distance around the outside of a circle is known as the circumference, or c

OPPOSITE: Geometric birch plywood treads ready to be fitted to steelwork.

Calculating Circumference and Area

CIRCUMFERENCE

Commonly you may need to calculate the circumference of a circular former in order to work out the size the cladding needs to be cut to, for example. To do this the formula is as follows:

$c = \pi d$
Or $c = 2\pi r$

So as an example, if I wanted to know the circumference of a circular former with a diameter of 300mm, I would type $\pi \times 300 =$ into my calculator to get a result of 942.477796076938. Given the radius instead, in this case 150mm, I would type $2 \times \pi \times 150 =$ for the same result.

Either way, I would probably round this up to the nearest half (0.5) millimetre for practical use, so the circumference would be 942.5mm.

AREA

To calculate the area of a circle, or A – that Is, the amount of space it occupies – the formula is   For this you can use the X^2 or 'square' function on your calculator. So as an example, if I wanted to know the square meterage of the top of a revolve that had a radius of 2.44m in order to work out roughly how many sheets of plywood I would need to cover it, I would type the following:

$2.44 \; X^2 \times \pi =$

Bisecting an Angle

Bisecting an angle is essential when dealing with mitred faces such as fitting moulding or making display plinths: it means that the length of the faces of the cuts you make will be the same size, resulting in a neat, seamless join.

There are a couple of techniques to use to achieve this: the first is for bench use using the material, the second a geometric method for draughting.

Method 1

Step 1: In the image we see the corner where two sides of an MDF plate meet. If we wanted to work out which angle to cut on the ends of two battens so that the corner was mitred, the simplest way of doing this would be draw two lines, parallel to the edges of the plate, using the batten as a guide set flush to the edge. Joining the point where these two lines cross with the corner of the plate will bisect the angle.

Step 2: Now an adjustable bevel can be set to this angle, and then taken to the mitre saw to set the correct angle for cutting. Remember that most mitre and table saws have a limit of 45–50 degrees, so bear that in mind when designing your scenery and planning your cuts to ensure they are achievable with the tools you have. You can always cut acute angles by hand, of course!

Method 2

The second method uses trammels to bisect the angle. As you can see in the image, we have two lines meeting forming the angle we want to bisect, as you might find on a footprint or technical drawing.

Step 1: Set the trammels apart – any distance is fine, just make it relative to the size of the markout. Here I have gone for 200mm. Now place the point of the trammels on the point where the lines meet, and draw an arc crossing each line.

45-degree mitres are quite straightforward, but bisecting angles accurately is essential when fitting moulding to more complex shapes.

Step 1.

Step 2.

Step 1.

Step 2.

Step 2: Move the trammel point to each of these new points and draw two more arcs, this time crossing inside the angle. Drawing a line between this central point and the point where the lines meet will bisect the angle perfectly.

TRIANGLES

One of the most common shapes required during the construction process is the triangle. It is a very useful form for all kinds of different applications, so it is important to have a good grasp of the basic principles relating to it.

Triangles obviously have three sides, so the three angles making up the triangle will always add up to 180 degrees, although the angles will not always be the same. There are several types you will encounter, as shown in the diagrams:

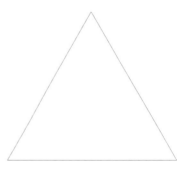

A triangle with the same angles in each corner, as well as the same length sides, is known as an *equilateral* triangle. Each interior angle is 60 degrees, so all three add up to 180 degrees.

A triangle with two matching angles and two matching sides is known as an *isosceles* triangle.

In a *scalene* triangle, each side is a different length.

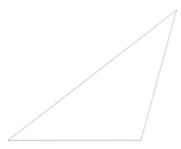

An *obtuse* triangle has one angle that is greater than 90 degrees.

In an *acute* triangle, all three angles are acute: they are all less than 90 degrees.

The *right-angle* triangle is the king of them all in construction terms. The fact that it features one square corner at 90 degrees, and the fact that the lengths of all the sides, as well as the interior angles, can be calculated using a straightforward method, means it is a very useful tool for laying out footprints, and checking for squareness.

PYTHAGORAS' THEORUM

Pythagoras' theorum states that: 'In any right-angled triangle, the square of the hypotenuse is equal to the sum of the squares on the other two sides.'

As a formula this theorem is expressed as: A2 + B2 = C2

This means that given the length of any two sides of the triangle, we can calculate the length of the remaining side.

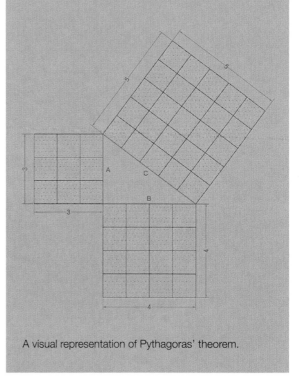

A visual representation of Pythagoras' theorem.

The key formula behind the right-angle triangle is known as 'Pythagoras' Theorum'.

As an example, let's imagine we are building a French brace, and want to calculate the length of the long angled timber, the hypotenuse of our triangle. Length A, 1,970mm, is the vertical height that the timber must reach from the top edge of the foot, while length B, 630mm, is the horizontal position from the stile to where it meets the foot of the brace. We want to calculate the length of C from

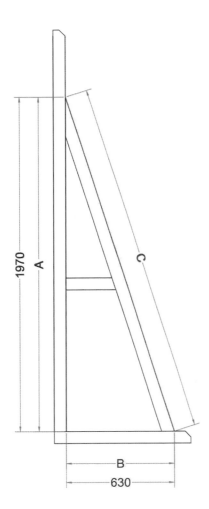

French brace exercise.

Technique 3-4-5

Probably the most useful 'shorthand' version of Pythagoras is a technique known as '3-4-5'. This ratio is an example of a right-angled triangle where all three sides are whole numbers, and therefore is a really useful way of checking for squareness over a large area or laying out large shapes. (See the previous section for a visual representation of this.) By measuring 3m on one axis and 4m on the other, the hypotenuse should be 5m if the two axes are perfectly perpendicular. These three numbers can be multiplied up for longer lengths if required, as their relationship remains the same. An example of using this might be for marking out a centre line perpendicular to the front edge of the stage, for example.

Step 1: Find the centre point on the edge of the stage, which we will call A, and measure 3m along the edge and mark another point, which we will call B.

point to point. So given that $A^2 + B^2 = C^2$ we know that $1,970^2 + 630^2 = C^2$. So on the calculator type:

$1,970 \ X^2 + 630 \ X^2 = 4,277,800$

Now we need to find the square root of this number by typing the $\sqrt{}$ symbol on the calculator. This gives a result of 2068.284313144593, so we will call the length of C: 2,068mm. I have rounded this down to the nearest millimetre for practical use, as I would struggle to find .28 of a millimetre on my tape measure, but the margin of error you work to is entirely dependent on what you are making – engineers work to very precise tolerances, for example, while carpenters have more fun!

Step 1.

Step 2.

Step 3.

Step 4.

Step 2: Set up trammels to 4m, and draw an arc heading upstage using the centre point A as the centre of the arc.

Step 3: Now set the trammels to 5m and draw another arc, starting at point B, so that it crosses the first. This is point C.

Step 4: Now snap a chalk line between point A and point C, and you will find that it is perfectly square (perpendicular) to the edge of the stage.

Step 3.

this case the base), allowing for the thickness of the bottom rail. An easy way to achieve this is to place an offcut of 3 × 1in PAR on the end of the stiles, and hook the tape measure on this to give an accurate reading. Mark the position of the centre of the rails using a combination or set square to mark across both stiles.

STEP 4

Next, measure either side of each centre line to indicate the thickness of each rail. In this case our rails are 22mm, so measure 11mm either side, and transfer these lines on to the edge of both stiles. Unclamp the stiles, and then, using a square, transfer the position of these lines on to the face side of each stile, which will help to ensure that the rails are fixed square to the frame. Do the same with the centre lines, but on to the back of the stiles. This will give us a good guide as to where to put fixings.

Step 4.

Step 5

Lay out the parts for the frame to double check that everything has been cut and marked out correctly. Begin assembling the frame, starting with the central rail and one of the stiles. This will generally help to support the frame on the bench as more pieces are fitted. Apply glue to the end of the rail, and line it up with the mark-out on the stile, ensuring that it is flush and square. Using the brad nailer, fix the rail in position, remembering to leave space for the screw holes to be drilled. Repeat the process for the remaining rails, and then the second stile can be fixed in the same manner.

Step 6

With the stiles and rails positioned, it is now advisable to fix them securely with screws to pull the joints tight before fitting the top and bottom rails. With the countersink drill bit fitted to the driver, drill two pilot holes into each joint, avoiding the brad nails and ensuring the holes are far enough away from the edge to avoid splitting. Now work around the frame, securing each rail with screws. Be sure to sink each screw head just below the surface of the timber, and ensure the joints are still flush.

Step 7

Now the top and bottom rail can be fitted in the same manner as the parts in the previous step; it is crucial that the outside corners of the joint are perfectly flush, as any discrepancies here will be very noticeable on the finished flat. With the main components fixed, the supporting blocks can be glued and fitted to each joint. Be sure to use shorter screws than on the main joints, otherwise they will stick through the frame.

Step 8

The frame can now be tidied up and finished, ready to be clad. The first thing to do is to wipe any excess glue off the joints using a damp cloth before it dries. Next, using a smoothing plane, take down any areas around the shoulders of the joints which are uneven. Decide which side of the frame is the back (usually the side which looks nicer, as the

Step 5.

Step 6.

Step 7.

add 60mm to the finished width of the frame, so that the head and sill (A) in this case would measure 1,360mm. On the cutting list express this as:

A: 2@ 1,360mm H&S ('head and sill')

STEP 3

To calculate the length of the stiles, we need to take two things into account: the face width of the timber, and the required length of the tenons. Check the face width of the timber stock: in this case the timber is 70mm wide, but remember that this can vary from one supplier to another, and even from batch to batch, so always check first and adjust your calculations accordingly. Each tenon will be machined at 50mm long, which is an appropriate size relative to the timber dimensions.

Therefore to calculate the length of the stiles, simply subtract the combined length of the two tenons (100mm) from the combined face width of the head and sill (140mm), giving a difference of 40mm. We then subtract this from the total height of the flat, therefore in this case the length of B is simply 4,380–40. On the cutting list, express this as:

B: 2@ 4,340mm 2T (two tenons)

STEP 4

We should apply shoulder length (*see* box opposite) to the calculations for the toggle rails C. As mentioned previously, these will be made from the thinner 22mm timber, so be sure to note this on the cutting list.

Calculate the shoulder length of C. To do this, subtract not only the combined width of the stiles, but also the width of the toggle shoes, so in this case 1,300mm–280mm = 1,020mm. Now add 100mm for the tenons to find the total length of C, which would be 1,120mm. On the cutting list, express this as:

C: 3@ 1,120mm 2T (two tenons)

The shorthand calculation (the same principle as Step 3) for this would be to subtract 180mm

SHOULDER LENGTH

The shoulder length is taken between the inside edges (shoulders) of the tenons along a rail or stile. Subtracting the combined face width of the stiles (2 × 70mm = 140mm) from the finished height of the frame (4,380mm) allows us to work out the 'shoulder length' of the stiles, so in this case the shoulder length of B would be 4,240mm. Now the length of the two tenons, in this case 50mm each, can be added back on to the shoulder length to give us the total length of 4,340mm for piece B. This is a slightly different, more thorough method for calculating the length of tenoned pieces, and understanding the use of shoulder length is particularly important when working with toggle shoes and flats with multiple stiles where tenon length may vary. It is also a good way of double checking measurements on cutting lists.

The shoulder length is measured between the inside edges of the two tenons.

from the total width of the frame to get the same result. The finished cutting list should look like this:

3 × 1.25in (70 × 28mm) PAR
A: 2 @ 1,360mm H&S ('head and sill')
B: 2 @ 4,340mm 2T (stiles)
3 × 1in (70 × 22mm) PAR
C: 3 @ 1,120mm 2T (toggle rails)

Toggle shoes: 6 @ 450mm
Corner braces: 4 @ 800mm

Step 5

Cut the pieces to length, working through the list in descending size order. As each piece is cut, mark the corresponding letter or dimension in the centre. Pieces requiring tenons should be marked with a cross at each end. Mark a face side and face edge on to each length, usually on the side with the fewest knots or imperfections.

Step 6

Machine the tenons on to the rails and stiles, not forgetting to adjust the height of the tenoner to account for the thickness of the timber being cut. Ensure that the pieces are consistently machined by placing every piece face side up in the tenoner.

Step 7

Now we can mark out the stiles for toggle rail positions, and the head and sill ready for mortising. Start by clamping the two stiles together face to face, with the face edges up. Fit your mark-out spacer to the end of one of the stiles to give you an accurate point to hook the tape measure on to. This will simulate the true length of the joint when the frame is assembled.

Mark the centre position of each rail on to the edge of the stiles, and use your combination square to transfer these lines on to both stiles. Now measure and mark out the face width of the rails either side of centre (35mm each way in this case). Unclamp the stiles and lay them face side down. Now transfer the rail positions on to the rear face of the stiles to aid assembly.

Now, using either your combination square or haunch marker, mark the haunches on to each tenon, ensuring they are on the opposite side to the rail mark-out. These can now be cut using a tenon saw, and then the stiles put to one side.

Step 8

Clamp the head and sill together face to face, with both face edges up. Measure 30mm in from one

Step 5.

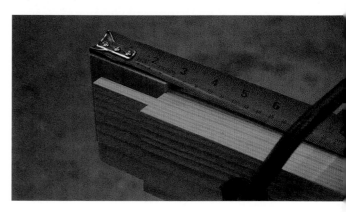

Step 7a: Marking out the stiles with the spacer fitted.

MARK-OUT SPACER

Machine a slightly oversized mortise into an offcut of your stock material, and cut one side off it. This will make a handy spacer, which can easily be fitted on to the tenons while marking out to represent the finished dimensions of the frame.

A simple mark-out spacer made from an offcut of timber.

Marking Haunches

To accurately mark the haunches on to your corner tenons, proceed as follows:

Step 1: Find the depth of the haunch by loosening the ruler on your combination square and sitting it into the haunch of a mortise (make sure it has been cleared out with a chisel), with the stock of the square sat flat on the edge of the timber. Lock off the ruler, remove it from the mortise, and make a note of the depth, in this case 20mm.

Step 2: We use the width of the combination square ruler to dictate the width of the haunch, so loosen the ruler again, this time placing the stock of the square on to the end of the tenon you wish to use, and push the ruler up to the shoulder. Mark a pencil line next to the ruler to mark the depth, 20mm in this case. This is the step that most people forget when starting out: the measured depth is what we want to leave on the tenon, not remove.

Step 3: Reset the ruler so that the end is in line with the pencil mark, and mark along the end and side of it to indicate the area that needs to be cut out, and hatch it out. You can now leave the combination square set, and simply draw around it on the remaining corners – there is no need to measure it out every time.

Step 4: You can also make a simple haunch marker using a piece of polycarbonate, by cutting it to the size of a haunched tenon; it is a useful thing to have in your tool kit if you use the same set-up regularly.

Step 1.

Step 2.

Step 4.

Step 3.

Step 7b: Transferring the mark-out on to the face makes positioning the toggle rails easy.

Step 7c: Use a tenon saw to cut the haunches, cutting across the grain first.

end of the pieces, and draw a line across with your square. From this line measure the total width of the frame, in this case 1,300mm, and mark another line across. Now the width of each mortise can be added in: double check the face width of the stiles (70mm in this case) to ensure an accurate mark-out, and mark 70mm in towards the centre from the outer lines.

The mortises should now be divided up to indicate the two different depths required. Use the width of the ruler on the combination square, and mark in the width of the haunch, remembering that it is placed towards the outside of the mortise next to the horn. Write the letter 'H' in the haunches, and mark a centre line through the full depth part of the mortise. Now hatch out the horns. The head and sill should look as shown in the image.

Step 9

Now the mortises can be machined on the head and sill. As mentioned in Chapter 3, Key Construction Skills, it is good practice to machine the two ends of each mortise first, then machine the

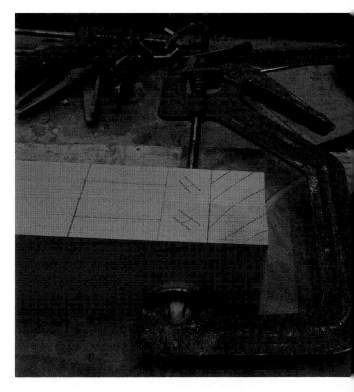

Step 8.

Step 9a: These mortises have been machined to two different depths to allow for the haunches.

Step 9b: Use a narrow chisel to remove any remaining debris from the mortise.

remaining area. Deliberately machining over the end on to the horn is also recommended, as this will make frame assembly slightly easier, as well as ensuring that the corner is nice and crisp when the horn is removed. With the mortises cut, they should be cleaned out using a 6mm chisel to remove any debris still present, which could prevent the joints from fitting correctly. Lay the timber on its side next to the edge of the bench to do this, as this will prevent chips from falling back into the joint.

STEP 10

The toggle rails can now be assembled and fixed. For convenience, the rails C can all be arrissed at this point with a block plane – be sure to arris only the long edges, and not the shoulders of the tenons. Loosely fit the toggle shoes to both ends of all the rails, using the wooden mallet to tap them flush at the shoulder. Any with persistent gaps should be cleaned out some more with the chisel. Lay out two sash cramps on the bench, and screw them down parallel to each other. Each toggle rail in turn

Step 10a: The toggle rail is clamped and ready for fixing with pegs.

should be clamped tight, adjusting the clamps to ensure that each joint is flush and square.

Now drill two 8mm holes, offset from each other to prevent splitting, through both toggle shoes. It is advisable to use a piece of scrap underneath when drilling to prevent breakout. With the holes drilled, hammer two wooden pegs into each shoe, ensuring there is space underneath for the pegs to come through.

STEP 11

Cut the pegs off flush on both sides of each toggle rail. The face of each joint can now be cleaned up using a sander to ensure that the joints are flush. On a task such as this with multiple pieces to clean up, it is advisable to screw some timber blocks to the bench to aid sanding. These will prevent the toggle rails from moving, and will eliminate the need to keep clamping them down, which can get a bit tedious if working on a large batch!

The final stage of preparing the toggle rails is to transfer the position of the outer edge of each rail on to the shoes. Ensure this is done on the face side, using the combination square. This will make positioning the rails simpler when the frame is assembled. Although it is possible to use the centre of the toggle shoes to position rails, this method is not always as accurate due to the fact that the position of the mortise on each shoe may vary slightly. With all the toggle rails marked up, assembly of the main frame can begin.

Assembly

STEP 1

Lay one of the stiles face down on the bench so that the mark-out is visible, as we want to attach all the toggle rails to this stile with the shoes set flush to the back. Begin by clamping the shoe to the stile using two speed clamps, using a mallet to adjust the position until the marks on the rail match the stile, and they are level. Once positioned, screw the shoe to the stile using two

Step 10b: Offset the pegs to reduce the risk of splitting the joint.

Step 11a: Sanding the joints flush.

Step 11b: Transfer the width of the rail on to the edge of the shoe for straightforward fitting.

5 × 70mm screws, and move on to the next rail. Now the second stile can be attached using the method above, starting at one end and attaching each rail in turn.

STEP 2

Next fit the head and sill to the frame. The joints must be pulled tight using sash cramps, either fitted with extensions to reach the ends of the frame or to the nearest rail. Be careful if you decide to do the latter, as too much pressure on the toggle rail will cause the shoe to slip. Having to apply a lot of pressure can be a sign that the joint needs some further cleaning out. Using the clamps and a tape measure, tweak the frame square, as shown in the key skills section.

With the frame still clamped to keep things tight, drill one central hole into each joint and hammer a peg through. Remove the clamps, and double check the diagonal measurements to ensure the frame is square. Assuming it is correct, a second peg can be fixed into each joint, offset from the first to avoid splitting. Hammering pegs into the joints away from the bench or trestles causes the frame to rattle around, so use the wooden mallet to support the joint underneath while you do this if there is no way to move the trestle closer.

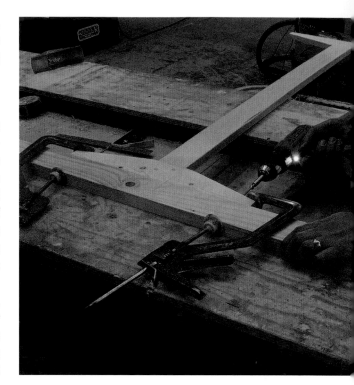
Step 1.

STEP 3

As this frame is intended to be covered with fabric rather than a hard sheet material, corner braces should be fitted in the corners to keep the flat square. As can be seen on the drawing they measure 800mm and require a 45-degree angle to be cut on each end. These should be glued and screwed in position, set to the back of the frame.

STEP 4

The flat can now be cleaned up. The pegs can be trimmed flush using a handsaw or Japanese-style pull saw. Next, cut the horns neatly off the corners using a handsaw, having scribed a line across the face as a guide. Now the frame requires sanding and arrising on both sides using a block plane, and the flat is complete and ready to be covered.

Step 2.

Step 3.

Step 4a: Cut the pegs off flush with the face using a handsaw.

Step 4b: Mark a line square across the horn to ensure an accurate cut.

The finished frame is ready for covering.

CLADDING

Cladding refers to the process of covering a flat or scenic framework with a sheet material to make it appear solid. This is usually done using plywood or MDF, but a variety of materials such as twin wall, sheet metal or polycarbonate can be used for this purpose.

The following method can be used for most cladding applications regardless of the style of the flat, the only difference perhaps being that the fixings and trimming methods will vary depending on your choice of sheet material – but the process remains very similar. The overriding principle here is to use the sheet material to set the frame square and flush using the 'factory edges' of the board.

We are going to look at cladding this timber flat with 4mm class 1 plywood. The ply will be glued and stapled to the frame for a neat finish.

You will need the following:

- Combination square and pencil
- Tape measure
- 4mm plywood or similar
- PVA wood glue
- Narrow Crown Air Stapler w/16mm staples
- Carpenter's pliers
- Router or laminate trimmer with bearing guide flush cutter
- Block plane
- 120g glass paper and block

Method

STEP 1

The first step is to double check the dimensions of your flat. In many cases it is best to deliberately cut the ply oversize to achieve the neatest possible finish. If you cut your sheet material to the exact size of the flat, you will find it difficult to get all of the edges of your frame to be perfectly flush, and will risk having small gaps where flats butt together. By oversizing the sheet slightly you can trim the cladding flush to the frame with a router. I usually add 5–10mm to both the width and the height of the cladding sheet for this purpose. Be sure to decide which way you want the grain of the plywood to run before cutting. In this case, the flat is 8 × 4ft (2,440 × 1,220mm), which is also the size of a standard sheet of plywood, so the sheet will only need a minor trim for neat edges.

STEP 2

Next we want to ensure that we know where the internal structure of the frame is when we come to fix the cladding, so mark the position of any rails to the outside edge of the frame using your square if it is not already clear from where the screw heads are. When working on flats with built-in features or extra support (such as a window or diagonal bracing) it is usually worth tracing the outline of the framework on to the cladding prior to gluing. Just be sure to do this with the sheet underneath the frame, rather than laid on top so as not to create a mirror image! By doing this it will be much easier to find all of the rails with the stapler, and will mean you don't have to waste time digging out rogue staples that have missed the frame.

STEP 3

We want to apply glue to the face of the frame, including all the internal rails. With flats built 'on face', use a zig-zag pattern to ensure that the glue spreads across the full width of the timber, and then lay your sheet on to the frame. Avoid dragging the sheet across the glue as this will create a mess on the back. When gluing up your frame, apply glue to the furthest edges first and work towards you to avoid leaning on it and spoiling your lovely T shirt!

Choose a corner of the frame to start from, and position the sheet so that it is perfectly flush to that corner. Don't worry if the frame and cladding run in and out of square around the edge at this point – the important thing is that your starting corner (and around 100mm either side) is perfectly flush. Some sheet materials such as class 1 plywood have an ink identification stamp on one side. Ensure that this remains on the back as the ink can, in some cases, bleed through paintwork which can be very annoying for scenic artists!

STEP 4

With your plywood sheet positioned, fix the corner down with the stapler. Put three staples either side of the corner equally positioned around 20mm apart, and 10mm from the edge. This will give the corner plenty of hold as you start to move the sheet around to get it flush. Where possible the head of the staple should be parallel to the grain of the plywood for the best finish. The first side to fix is the short edge of the frame adjacent to your starting corner. This will give you the maximum amount of leverage when it comes to squaring up. With this in mind, work your way along the edge of the frame (always working from you starting corner), moving the sheet until it is flush with the edge and stapling down at 70mm (3in) intervals until you reach the next corner.

Step 3.

Step 4.

STEP 5

Repeat the process, this time fixing the long edge of the frame. Notice how it is possible to move the frame dramatically in and out of square as you work your way up to enable you to get the edge as flush as possible. On larger flats it is advisable to work with someone else to move the cladding into position whilst you fix. The frame is now square to the cladding and you can fix the remaining two edges. The cladding will be over-sailing the edges on these two sides, so mark the overhang by using your combination square as a depth gauge and running it along the face with a pencil. Arguably you could probably just guess where the edge of the frame is, but in the interest of best practice and the neatest finish it is worth spending the time marking out. With the edges marked, continue stapling the sheet down until you reach the last corner.

STEP 6

Transfer the position of any rails on to the face of the cladding if you haven't already done so, and staple down on to all the internal structure of the flat to ensure a nice smooth surface. 'On face' flats require a second line of staples on each rail to achieve the flattest finish. Skipping this step, or only partially completing it, will potentially lead to the cladding bulging when painted, or being more susceptible to damage. With the sheet fixed down, crouch down so that the flat is at eye level and check that all the staples have sunk just below the surface of the ply. Any left sticking out should be hammered flush or pulled out with carpenter's pliers. This should be done prior to trimming, as the base of the router will catch on any protrusions.

STEP 7

Set up your router so that the bearing is running on the timber frame at a depth avoiding any screw heads or knot holes. Working in an anticlockwise direction (the opposite the direction in which the cutter is spinning) around the outside of the flat, trim off the overhanging ply with the router. Use

Step 5.

Step 6.

your block plane or glass paper to take the edge back slightly, but don't get carried away, as it is important to retain a crisp edge to ensure neat joins once multiple flats are butted together in situ.

The flat is now ready to be painted!

COVERING WITH CANVAS

Flats can be covered with soft fabrics such as canvas for scenery, as well as bolton and serge for masking purposes. Canvas flats are generally covered on the face with the canvas glued to the frame. New canvas is quite stretchy and will tighten up as it is painted, so care should be taken not to over-stretch it at first or you may find that your flats will warp. Masking flats will generally have their serge pulled tight and wrapped on to the back to mask the edges. Sometimes pre-painted canvases may need to be transferred on to new timber frames, in which case it is usually better to wrap the canvas on to the back also. Painted canvas is much stiffer and requires more work to get even tension into, so investing in a pair of canvas pliers is advisable!

Traditional mortise-and-tenon flats are ideal for covering in this way; flats built on edge are less suitable due to the fact that the tension in the canvas will cause the outside of the frame to bow. The canvas is gently stretched across the frame and temporarily held with staples. The edges are then glued down and the staples removed. Once primed, the canvas will be nicely taught, giving a very smooth, flat surface on which to paint. The following section describes how to go about covering a flat with new canvas.

You will need the following:

- A mortise-and-tenon or 'face on' flat
- Scenic canvas
- A wide-mouth staple gun
- 8mm staples
- Carpenter's pliers (pincers)
- PVA glue
- Stanley knife

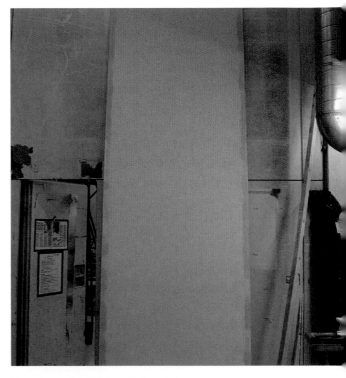

A canvas-covered flat ready for priming.

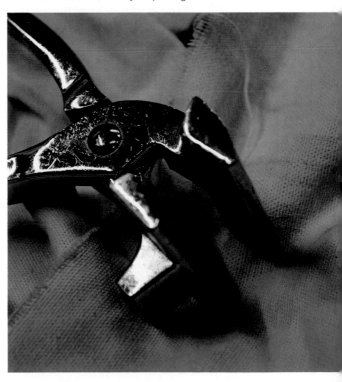

Canvas pliers make re-stretching cloths much easier.

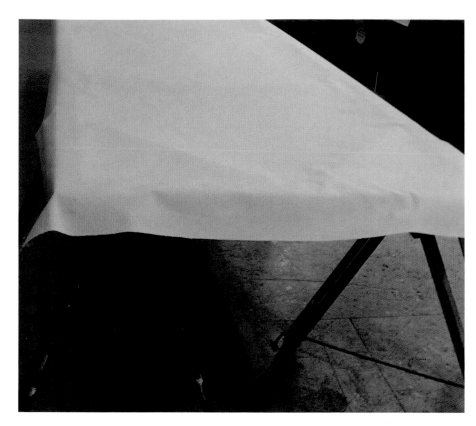

Step 1.

Method

STEP 1

Lay the flat out on trestles or a workbench, face side up, and unfold the canvas over the frame. Allow around 100mm of excess on each side, and trim the canvas to size with a sharp Stanley knife. It is always worth using a fresh blade otherwise the canvas will fray quite badly.

STEP 2

The next step is to temporarily stretch the canvas over the open area of the frame. Begin in the centre of one side, and put in three staples spaced about 30mm apart, set at a slight angle by tilting the stapler so that they can be removed easily later. The staples should be set towards the inside edge on the face of the frame, to allow us to glue the canvas down. Now go round to the opposite

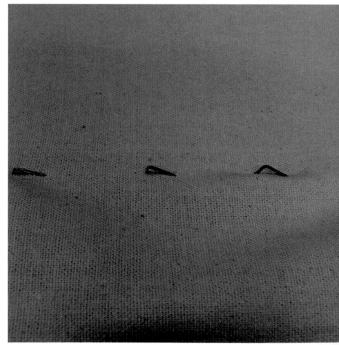

Step 2.

Step 3.

side and repeat, pulling the canvas fairly tight but not overly so.

STEP 3

Now repeat Step 2, except this time on the top and bottom edges. You should see that there is a cross shape where the canvas is taught. Now working from the centre out, work your way along each edge fixing staples in threes, mirroring each side to ensure equal tension. The reason for doing this is that any creases can then be chased out to the corners and won't become trapped in the middle of the frame. The frame should now have the canvas pulled fairly flat over the surface, fixed evenly all round.

STEP 4

Fold in the loose canvas on the edges to expose the timber frame beneath. Work your way along each edge with PVA glue, slightly watered down

Step 4.

Piece A will be the full height of the brace, minus the width of the sill. So this would be 2,440mm–70mm, expressed as A) 1@2,370 on the cutting list.

Piece B will simply be the full length of the sill of the brace. This will expressed as B) 1@800 on the cutting list.

Piece C has been dimensioned on this drawing, so in this case I will add 150mm to the length to make marking out and cutting the angles easier. If we didn't know the length of 'C' – for example, if we were designing the brace from scratch – it could easily be calculated using Pythagoras theorem $A^2 + B^2 = C^2$ to give us an accurate dimension. For more information on this, refer to Chapter 4, the maths reference section.

Piece D has also been dimensioned. For the purposes of cutting, adding 20mm to the length which will help marking out the angles.

It is rare to need only one brace, so the first is a good template from which to build more frames. When this is the case, once all the pieces are cut to size with the angles marked out accurately, subsequent pieces can be marked out and cut without the need for laying out each one, using the first set as a template to save time and ensure consistency.

STEP 2

With all the timber cut, the next step is to make the plywood plates that will support the butt joints of the frame. The shape and dimensions for these can be seen on the diagram. Note that the grain should be running in a direction where it offers maximum support to the joint. Unlike most flats, which would only be plated on one side, the brace has no cladding, so two sets of plates are required. With the plates cut, spend some time cleaning them up with a block plane and sandpaper to give a nice neat finish, and set them aside.

STEP 3

The next phase is to lay out all the pieces of the brace and mark out the angles for cutting. This is where having a solid workbench is useful, as

Step 2.

being able to clamp or nail pieces down is crucial. If working from trestles or on the floor, I would recommend laying down a sheet of plywood to use as a flat surface. Begin by clamping or screwing down 'A' and 'B' in position, ensuring that they meet each other at 90 degrees. If working on a bench, several battens may be screwed to the edge to provide a physical stop to butt the pieces up to; while if working on a board, a good method for ensuring squareness can be to use the factory edges of the sheet to line the pieces up flush to. This can be double checked with a roofing square.

With these pieces securely fixed down, measure and mark a line 100mm in from the toe end of the sill, and 400mm from the top of the stile. Also mark the position of 'D' on the stile and screw it down, again ensuring that it is perpendicular to the stile.

STEP 4

Now piece 'C', the diagonal, can be laid on top. This piece is oversized, so roughly split the difference in the overhang on each end, and line up the

Step 3.

Step 4.

top edge with the mark-out, and fix it down with clamps or nails.

Using a combination square, scribe lines up on to the edge of 'C' from 'A', 'B' and 'D', where it intersects them. Also scribe lines across the face of these pieces, as this will give us an accurate mark-out for cutting angles and assembling the brace.

All the pieces may now be taken up.

STEP 5

With the pieces provisionally marked out, set up a sliding bevel to the face angle of 'D', and use it to mark the face of 'C' where it meets the sill. Connect the marks at the top of 'C' using a straight-edge or saw blade to show the angle of the longer taper cut. The 45-degree angles can also be marked out on to the top of 'A' and toe of 'B'.

If you are making a batch of several braces, it is advisable to transfer the mark-out on to the rest of the pieces at this point: this is best done by clamping like pieces together on edge, and scribing across. This saves a lot of time and improves consistency compared with marking out each frame individually, and should be done prior to cutting angles to ensure accuracy.

Cut the angles into 'C' and 'D' either by hand, or with a jigsaw or mitre saw. The angle on the mitre saw can be set using the sliding bevel as a reference. Any rough surfaces should be planed smooth, and the pieces can be laid out on the bench ready for assembly.

STEP 6

Apply glue to the contact areas of each joint, and lay the frame together. The joints will be held together using corrugated fasteners known as wiggle pins. These should be fixed in pairs, centred over the butt joints, and toed in slightly to ensure a tight joint.

Start with the right-angle corner where A and B meet, then attach D. With these pieces held together, C can now be fixed in position. Start at the base, ensuring that the joints all correspond to the mark-out. The long taper at the top of C should be screwed in using 2 × 5 × 60 screws, drilled

Step 5.

Step 6a: Fixing with wiggle pins.

Step 6b: Screwing the taper into position.

and countersunk prior to fixing. This is the most crucial join, because it will dictate whether or not the frame is square, so double check before fixing.

STEP 7

The frame can now be cleaned up ready to have the ply plates fitted, so use a sander on each joint to ensure they are flush, and work around the edges, arrising with a block plane. Repeat on the other side of the frame.

The final stage is to secure the plates down. Glue each plate and lay it in position, ensuring that there is a 5mm gap between the edge of the frame and the plate, and either staple the plate down using narrow crown staples or 4 × 20mm screws. The brace is now complete, and is ready to get to work!

MODIFYING A FLAT

Many theatres and workshops have a stock of flats from previous productions, which can be easily

WORKSHOP 'A'-FRAME BRACE

A useful addition to any workshop is a set of 'A'-frame braces. These can be screwed or clamped to the edges of flats and other pieces to stand them up temporarily for painting, storage, and fitting.

These braces are simply cut out from a sheet of 18mm (3/4in) plywood using a circular saw and jigsaw. The edges of each have been arrised with a laminate trimmer fitted with a 45-degree chamfer bit. This can produce a neater finish on plywood than using a block plane, as the grain is running in several directions.

An 'A-frame' brace temporarily fitted to a flat in the workshop.

modified to suit your project, rather than building from scratch every time. In this example I want to reduce the height of the flat by 300mm.

Step 1: Lay the flat face up on the bench, and measure 300mm down from each corner. Use a straight-edge to draw a line across the face of the flat.

Step 1.

Step 2.

Step 3.

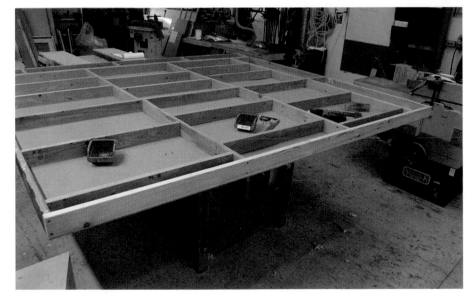

Step 4.

Step 2: Using a circular saw, cut off the top of the flat, through both the cladding and the frame. If the saw cut doesn't cut all the way through, then finish with a handsaw.

Step 3: Flip the flat over. We want to add in a new top rail, so the stiles will need to be cut back by the width of this piece, in this case 22mm. Mark this on to the stiles, and transfer the lines on to the sides of the timber as well, to help keep the

cut square. Now using a handsaw, cut the stiles back to the lines, being careful not to cut through the cladding.

Step 4: Cut a new top rail, and pilot the ends. Glue the ends of the stiles, as well as the edge of the top rail, and screw it into position. Flip the flat over again, and staple the cladding to the new top rail. Use a block plane or router to trim off any excess and neaten up the edge.

6
TREADS AND STAIRS

There are several different approaches to building treads and staircases that are useful for different situations on stage, but before any aesthetic considerations it is vital to make them structurally sound and be sure that they function correctly. Taking shortcuts here can lead to accidents, so consult the relevant guidelines relating to your project (the *ABTT Yellow Book* in the UK is a good start) to check what is best for you.

Project 4: Build a Set of Block Treads

A robustly reliable method of constructing smaller sets of treads (say, of between two and six steps) is to use the following block tread method. Being essentially a plywood box, block treads are stable, and can be set and struck

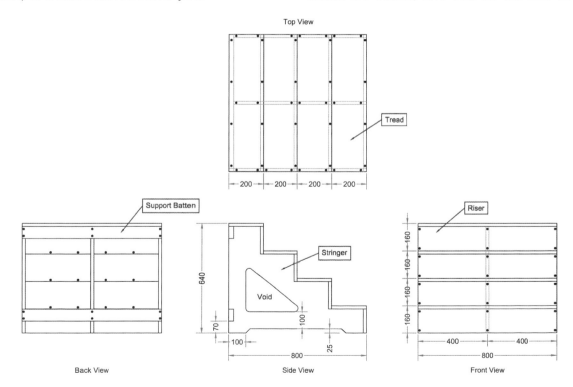

Project 4: Build a Set of Block Treads.

Step 6.

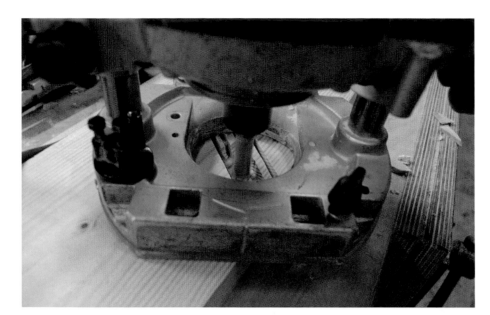

Step 7.

Step 8.

then do two passes at incremental depths. With the bulk of the waste removed, use your chisel to chase out the housing and leave a sharp, clean finish.

STEP 7

Next, drill two pilot holes through each housing for attaching the treads. Turn the strings over and use the countersink bit through the holes you have drilled. The treads are now ready for assembly. Dry fitting a tread into every housing is advisable, as it is much easier and neater to deal with discrepancies before any glue is involved!

Take the first string and sit it on the bench on edge. Apply glue evenly along the end of the first tread, and push it into the bottom housing. Now screw it in place, ensuring that it is seated correctly and pulled tight. Avoid sinking the screws too far into the string: the head should be just shy of the surface.

Now repeat this process with the top tread next to improve stability on the bench, and then attach the remaining treads. Attach the second string in the same manner: putting the assembly on edge on the floor may help to line everything up.

STEP 8

Now clean up the treads, taking back the front corners of each step, and arrising the entire piece all round. You may wish to add ties into the back of the treads to reduce the likelihood of the strings separating. These can be battens notched in, glued and screwed into the back edge of the stringers, usually one near the bottom and one near the top, but these aren't always necessary.

The final step is to mount any ironmongery to the treads for fitting them. These are designed to run up to a steel rostrum, so have peg plates fitted to the inside faces of the strings in the correct position to tighten them up to the deck.

7
DOORS AND WINDOWS

DOORS

Making a door in a scenic style as opposed to employing the traditional joinery approach serves several purposes. The first is that it is possible to scale up the design of the door quite easily without being limited to the available range of softwood stock. For example, a traditionally built panelled door may have a top rail at 150mm (6in) and a bottom rail of 225mm (9in), but if a set design requires this style of door but at twice normal size, not only will it be more difficult to find suitable timber, but it may also be prohibitively expensive to buy and then make the door.

OPPOSITE: *As You Like It*, RADA Jerwood Vanbrugh Theatre. Director: Michael Fentiman; designer: James Cotterill. © RADA

Doors come in many styles, shapes and sizes, and can be built to suit the specific requirements of the production.

fitting, to ensure that everything is in the right place when it is attached. This batten will help to prevent the flat from getting damaged when it is moved around.

Another way of doing this is effectively to build the travelling batten into the frame, and cut it out later. To do this, simply run the bottom rail (sill) across the full width of the flat, including the doorway. This will need to be cut out prior to the reveal being fitted, but it will make life a lot easier during initial assembly as the flat will happily sit on the bench without having to worry about it twisting, and it will be easier to square up.

Fitting Butt Hinges

The easiest way of fitting hinges accurately is to mark up the frame and the door at the same time. It is quite unusual to hang the door in its frame permanently until the fit-up, mainly so that the scenic artists can get to all the edges of the frame and reveals, and also because handling and manoeuvring the flats will be easier without the added weight of the door. However, the more prefitting that can be done prior to paint, the better, as the doorstop can be accurately fitted, the door correctly trimmed to fit the hole, and the hinges marked out, which will mean less fiddling about on stage.

Step 1

I would recommend standing up the flat containing the door frame and temporarily bracing it as if it were sat on stage. It is possible to fit the door with the frame laid flat, but it will not be easy!

As the door stop hasn't yet been fitted, be sure to clamp or brad a piece of timber or two to the inside of the frame to prevent the door from falling through the hole and making you feel like an idiot (trust me). Offer the door up to the aperture and check the fit: ideally there should be a 3mm gap between the top of the frame and the door, and the same at the sides. You might want to allow more clearance at the bottom of the door, especially if

there is carpet or texture to contend with, so a gap of anything up to about 10–15mm is acceptable. If the door needs to open out on to a floor that is considerably higher than in the closed position, then I would recommend using rising butt hinges that will lift the door up as it opens, otherwise there may be a huge gap under the door when it is in its closed position.

The door should be planed along its edges and trimmed with a circular saw if more material needs to be removed, until you are satisfied with the fit, at which point it can be aligned with the frame using packing to maintain the correct gaps.

Step 2

Next we will mark the position of the hinges on both the frame and door. Double check that you are working on the correct side of the door, as in the opposite side to the lock block. On manufactured doors the lock-block position is usually printed on the top edge of the door.

Measure down 150mm (6in) from the top of the doorway, and mark the edge of the door

Six from the top, nine from the bottom.

WHY AREN'T THE HINGES SPACED EQUALLY?

There are a couple of theories over the reason behind the traditional '6 & 9' hinge spacing. One is that this is purely for aesthetic reasons, and is to compensate for 'foreshortening' due to the fact that the top hinge is closer to eye level. If the hinges were equally spaced, they would actually appear to be unbalanced.

Another theory is that the hinges are spaced to avoid the tenon positions of a traditionally built door. The hinge would be cut in just below the 6in top rail, and just above the 9in bottom rail, to maximize the support for the weight of the door, and to avoid screwing directly through the shallow part of the stile and risk splitting the tenons within. Of course, this wouldn't make any difference on a hollow scenic door or modern moulded door, where the stiles run the full height of the door without any mortise-and-tenon joints.

In my opinion both are correct. I suspect that the traditional stock sizes for the top and bottom rails were originally selected to appear balanced once the door had been constructed, rather than for purely structural reasons, and the hinge positions naturally followed on from that.

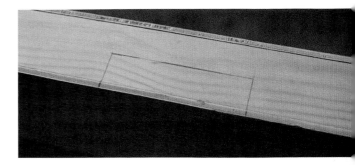

Step 3.

Step 4.

and frame. Measure up 230mm (9in) from the bottom of the door, and mark across here also. This indicates the outer edge of each hinge. Remove the door from the frame, and either clamp it on edge to the workbench, or preferably use a saddle

STEP 3

Measure the length of the hinges you are using: here I am using 3in butt hinges, which come in at 76mm across. Mark the length on to the edge of the door using a square, measuring in from the marks you made previously. Try to avoid drawing around the hinge itself, because tempting as it is, you will end up cutting out a wider area than necessary, and keeping the hinge perfectly square to the edge won't be easy. A compromise is to 'tick off' the length of the hinge at the edge, then

switch to a square. Next, measure the width of one leaf of the hinge, and using a combination square or a marking gauge, scribe a line between the length marks.

STEP 4

Next we need to mark the depth of the hinge accurately as a guide for paring out. The idea is for the top surface of the leaf of the hinge to be flush with the edge of the door when fitted. This is best done with a marking gauge, lining up the point of the single pin with one side of the hinge and the stock of the gauge with the other. With the gauge set, press the pin into each end of the marks, and then carefully score a line into the edge between the two points. With the width, length and depth of the hinge marked out, hatch out the area ready to be removed.

MAKE A SADDLE AND BLOCK

A saddle is basically just a large piece of timber with a notch cut out of it to sit the door in. The door is held in position with the wedge. The block is simply another piece of timber to rest the door on at the other end, cut to the same thickness as the recess in the saddle, to keep the edge of the door horizontal to the floor. These will make working on doors really easy, especially on stage. They are simple to make using offcuts such as timber bearers, and you will wonder how you ever did without them!

You will need the following:

3 × 2in (70 × 44mm) PAR or similar
Combination square
Pencil
Handsaw
Chisel
Optional:
Mitre saw
Jigsaw
Bandsaw

Step 1: First cut a length of timber at 610mm (24in), and put the piece up on the bench for marking out. Find the centre at 305mm, and scribe a line across. The saddle can be made to suit doors in a range of thicknesses due to the wedge shape. Most timber doors I work on are between 40 and 50mm thick, so I will mark out a width of 45mm, working 22.5mm either side of the centre line; obviously this size can be adjusted should you require it. Now transfer these lines down the side of the timber to a depth of 35mm.

Step 2: The next step is to make the wedge. I tend to cut the wedge out of the saddle timber, as the two can then easily be kept together for storage. The mark-out should look something like that shown in the diagram.

Step 3: Cut out the wedge either on the bandsaw or the jigsaw or by hand, and clean it up if required. Now lay the wedge to one side of the notch mark-out, so it overhangs the edge equally. Mark the angled edge of the wedge on to the

saddle, and transfer the lines down the sides to the depth of 35mm.

Step 4: The notch can now be cut out of the saddle. For convenience I have used a mitre saw here with a depth stop set to 35mm to make a series of relief cuts approximately 5mm apart, but these can obviously be cut by hand. Then using a chisel, clear the notch out to finish with a smooth surface for the door to sit in.

Step 5: Rip an offcut of timber to a suitable size for the block. In this case it will be 35mm thick to match the notch in the saddle. Now the door can be placed into the saddle, and the wedge tapped in to hold the door still; you may want to line the saddle with thin packing to protect the surface of the door, particularly if it has already been painted.

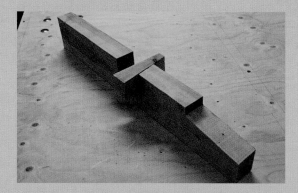

A saddle and block make it much easier to work on a door.

Step 1.

Step 2.

Step 3.

Step 4.

Step 5.

STEP 5

The next step is to remove the waste from your mark-out so that the hinge slots as perfectly as possible into the edge of the door. This can be done in a couple of different ways, either by hand using a chisel, or also using a small router (or laminate trimmer). In either case, the first step is to neatly mark around the shape of the cut-out using a wide chisel or Stanley knife and straight-edge. This will help to keep the finish neat, leaving the pencil line just visible, as well as the fit tight.

Now the waste can be removed. Using a chisel, place the bevel side down, and holding

Step 5a: Score out the outline of the hinge with a chisel first.

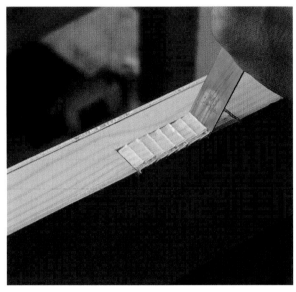

Step 5b: Walking the chisel back and removing the waste.

Step 5c: Keep checking the fit while making alterations to the door.

Step 6.

the chisel at an angle, begin cutting out the waste by walking the chisel back in approximately 10mm intervals, cutting down to the marked-out depth each time.

The waste can then be pared out by hand leaving a clean, level surface, and any edges can be neatened up. Using a router for this can be a huge time-saver, particularly when fitting multiple doors, and especially when used with a hinge jig. These can be purchased, but are also fairly straightforward to make, and essentially act as a guide for the router to cut out the correct shape almost perfectly every time. It is possible to achieve this without the use of a jig, but care should be taken to ensure that the cutter doesn't wander over the edges. The most important setting here is the depth of the cutter, so make sure to test the depth on a piece of scrap before you start on the actual door!

STEP 6

With the hinges let in, they can now be screwed into the edge of the door. Place the leaf of the hinge into the slot, and ensure that you are satisfied that everything is aligned correctly. Mark the position of the holes and remove the hinge from the door, then drill a small pilot hole in the centre of each of these marks.

Drilling off-centre will cause the hinge to shift when the screws are tightened due to the head of the screw centring on the countersink, so take your time with this. Try to use the screws supplied with the hinge, as the heads will match the shape of the countersink and should finish flush. If different screws are required, then increase the size of the countersink in the hinge to accommodate the screw head, otherwise the opposing heads will clash when the door closes, causing the door to bind, or damaging it. Tighten the screws and ensure that everything is still flush and true; if not, then remove the hinge and make small adjustments.

STEP 7

Now repeat Steps 3 to 5, but this time to let the hinges into the door frame. Use a spare hinge to check the fit, or one leaf if the hinge has a

Pack up the door to the correct height in the open position in order to fit the hinges accurately.

removable pin. Now mark and drill pilot holes, but only in the top position of each hinge. This will allow for some minor adjustment before committing to final fit should it be required.

Pack up the door in its open position with some scrap, or use a lever such as a board lifter under the door, and screw in the hinges using only the top holes. Remove the packing, and gently (there are only two screws in it!) close the door to check the fit. Make any adjustments to the hinge positions and door size as required, and then pilot and fit the remaining screws.

With the door packed up, the remaining pilot holes can be drilled into the frame.

Once the door is hung, the doorstop can be fitted.

Step 8

With the door in its closed position, mark the inside of the jamb with a pencil. The doorstop can now be fitted around the doorway using glue and brad nails. This will prevent the door from swinging through the frame and damaging the hinges, as well as preventing light leak through the doorway when closed. Note that it can be preferable to attach the doorstop once the latch has been fitted to ensure the closest fit. If required, the door can now be removed from the frame, although you might prefer to fit handles and latches with the door fitted. If using fixed-pin hinges, I would recommend leaving the hinges on the door, but removing the top screw, then closing the hinge and screwing it back in for safe transit.

Fitting a Tubular Latch and Handles

There are many different types of mortise latches suitable for use on stage doors. On timber doors traditional tubular latches are commonly used in conjunction with matching handles on a spindle for the most reliable functionality, but often roller catches and ball catches are used as an alternative where traditional door furniture is not being used. Whichever style is used, it will need to be fitted into the edge of the door, and the striking plate let into the frame. In the following example I am fitting a tubular latch to be used with lever handles; the basic principle is the same for catches, without having to drill out the spindle.

It is really important to select the correct size and type of tubular latch depending on the type of handles being used, as well as the proportions of the door itself. Lever handles are always positioned with the handle facing into the door, so generally a shorter latch can be used; however, a doorknob requires much more space around it to work well, so choosing a longer length is advisable to ensure that the members of the cast aren't constantly catching their hands on the door frame when turning the handle. Another consideration is that aesthetically, the position of the door handles should be relatively central to the lock stile of the door.

You will need the following:

- Tape measure
- Pencil
- Combination square
- Tubular latch and striking plate
- 22mm flat bit or auger bit
- 10mm Forstner bit
- Electrical tape
- Chisel
- Hammer
- Pilot drill bit and countersink

STEP 1

First decide on the height of the spindle for the handles. As a guide this should be between 900 and 1,000mm from the floor to be a comfortable height to use, and roughly central to the middle rail on the door, but the importance of this will depend on the design. Mark a line across the edge of the door at the chosen height, and transfer it on to the face and back of the door. Next measure the distance from the face of the latch to the centre

Step 1.

of the spindle hole, which is known as the 'back-set'. In this case it is 45mm as I am using a 63mm latch, so mark this on either side of the door on the centre line.

STEP 2

Measure the total length of the latch, in this case 63mm, and wrap a piece of electrical tape around the 22mm drill bit at this depth. Now find the centre on the edge of the door, and drill it out to the desired depth. A 22mm bit is a good size for the latch I am using, but you may need to increase this depending on the hardware you use.

STEP 3

Now using the 10mm bit, drill out the holes for the spindle. It is important to drill in from each side to ensure that the holes line up perfectly, and also to prevent unwanted break-out on the back of the door. Now clear out any waste in the mortise ready to fit the latch.

STEP 4

Push the latch into the hole, and rotate it so that it is sitting perfectly square to the door. Now mark the position of the screw holes with a pencil, and remove the latch. You will notice that the faceplate of the latch is countersunk and protrudes out of the back, so drill a small pilot hole and countersink for each of the screws to allow the latch to sit flat against the edge of the door. Push the latch back in, and carefully mark around the outside of it using a Stanley knife or sharp pencil.

STEP 5

The faceplate of the latch now needs to be let into the door so that it finishes flush. Using the chisel, cut around the outside shape of the plate first, then chop out the area evenly to the depth of the plate. Now countersink the pilot holes again, and refit the latch using the supplied screws. Ensure that the latch is the correct way round, so that the curved side of the latch tongue is facing the leading edge of the door. Double check that the spindle (normally supplied

Step 2.

Step 3.

Step 4.

Step 5.

Step 6.

with the handles) fits and rotates without catching the door.

STEP 6

At this point it is usually a good idea to fit the handles to the door for convenience. If the door is yet to be painted, you may want to use only minimal fixings to fit the handles to make it easier to take them on and off, and to avoid damage to the screw holes in the door. Ensure that the spindle is inserted through the door, and slide one of the handles on to it. Position the handle so that it is square to the door, and ensure that the spindle is still horizontal, then mark the screw positions with a pencil.

Drill pilot holes into the door, then fit the handle using the screws supplied. Remember that quite often the play will be set in a particular period of history, often in a time before posi-drive screws existed, and although this is a tiny detail, it is nevertheless one that designers and some eagle-eyed members of the audience might pick up on. Therefore use slotted screws, and tighten them by hand to avoid damaging the finish on the handles or door.

Fit the opposing handle, ensuring that the latch operates easily. If there is a lot of friction, move the handle slightly until it has freer movement, and reattach. If there are grub screws on the handles, then these can be tightened.

STEP 7

The striking plate can now be fitted to the frame, so close the door until the latch tongue touches the edge of it, and mark the position with a pencil. Measure the distance from the edge of the door to the square edge of the latch, and transfer this on to the frame also, to ensure that the latch catches with the door in the correct position. Open the door out of the way, and position the striking plate face down against the frame so it can sit flush. Carefully position it perfectly square, with the inside edge of the latch hole on your pencil line to set it the correct distance in. Now carefully draw around the plate and mark the screw holes.

stretched and stapled on to the back of the windows, but looks more 'realistic' than simply having an open frame.

Polycarbonate: This is the best choice for a realistic glass effect, but it needs to be carefully fitted. Thicknesses of around 8–10mm are ideal, as the panes will appear flat under lights. Using thinner materials will be cheaper, but cause very unrealistic reflections and distortion akin to a funhouse mirror, particularly if screwed into. The best method for fitting polycarbonate is to fit it into the frame in a rebate so there is no direct pressure on it. It can pick up scratches, so clean it with caution! It is also available in a wide range of colours and opacities to suit most designs.

Sugar glass: I would certainly not advise using sugar glass in every window on set, but it can be used to great practical effect if a window pane needs to shatter during a performance, as it is very safe to use and breaks quite realistically.

Project 7: Build a Georgian-Style Window

This project demonstrates how to go about making a window with glazing bars for stage use. It could be used as part of a sash or case-ment-style window, and the design can easily be adapted to suit different shapes. The purpose of this exercise is to use small stub mortise-and-tenon and half-lap joints, and to learn how to construct the glazing bars so it appears as if the window has been traditionally constructed in timber. Arguably you could simply CNC a window like this out of a sheet of ply – but where is the fun in that? In all seriousness, though, if you have a lot of windows to make where the finish is not so crucial that might be a good option! Be sure to read through the mortise-and-tenon flat project in Chapter 5 before attempting this project.

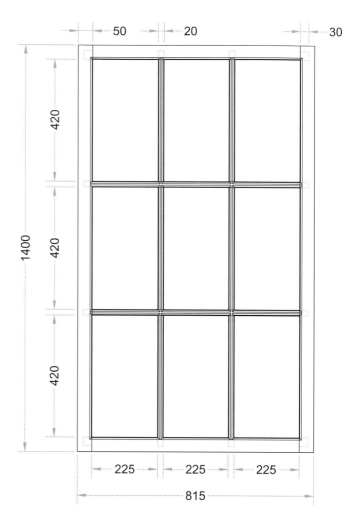

Project 7: Build a Georgian-Style Window.

You will need the following:

- Tape measure
- Pencil
- Combination square
- 50 × 28mm PAR timber
- 48 × 28mm PAR timber
- 20 × 28mm PAR timber
- PVA glue
- 8mm clear polycarbonate
- Router or laminate trimmer with ¼in shank bearing guide ovolo cutter

STEP 1

Study the drawing with a view to writing a cutting list. The first thing you will notice is that the stiles of the window frame run full height and the top and bottom rail are set in, which is the opposite way round to a mortise-and-tenon flat. We are going to tenon the glazing bars as well as the corners, however we won't use haunches on the corners in this case. Instead we are going to use stub corner joints, where the width of the tenon is cut down towards the outside of the joint. This method is better suited to the smaller size of stock we are working with than using haunches.

You will also see that the glazing bars cross each other by way of a half-lap joint and joint the outer frame with a mortise and tenon, so bear this in mind when writing the cutting list, as the bars are not all separate pieces.

As shown on the drawing, the tenons are all 30mm long, so set up your tenoner with this in mind. We are also working with thicker 28mm stock, so ensure that the tenons are set centrally on your test piece.

Label the component parts of the frame with a letter, in size order. As we are not using haunches there is no need to add on material for horns. To calculate the lengths of the glazing bars and top

and bottom rails, simply subtract 40mm from the height and width of the frames, due to the fact that the stock is 50mm wide, and the tenons are 30mm. With this in mind your cutting list should look something like this:

50 × 28mm PAR
A: 2 @ 1,400mm
B: 2 @ 780mm 2T
20 × 28mm PAR
C: 2 @ 1,360mm 2T
D: 2 @ 780mm 2T

Cut the pieces to length in descending size order, remembering to label the letters and mark face sides and edges as you go. Any pieces requiring tenons should be marked with an X at each end.

STEP 2

The next step is to machine the tenons, and then mark out the mortises. Clamp the top and bottom rail together, face sides in, face edges up, and make sure that the tenons are perfectly aligned in the clamps. Hook your tape over the shoulder of the tenon, and mark out the position of the vertical glazing bars (C) by marking the centre and measuring 10mm either side to indicate the width of the

Step 2a: Marking out the mortises.

Step 2b: The stub corners are marked out.

mortise. Draw a freehand line down the centre of the mortise and put the rails aside.

Repeat the process, this time with the stiles (A), again ensuring that the ends are perfectly aligned. Mark the glazing-bar mortises as well as the wider 50mm mortises at each end for the top and bottom rail. As we are using stub corners, the first 20mm at each end is not machined, and the mortise is only 30mm wide, so mark this on to the stiles.

Rather than mark out the half laps on the glazing bars separately, transfer the layout from the outer frame on to these pieces by clamping them together, this time with the bars face up, but the outer frame pieces on edge. Mark an X in each half-lap area.

STEP 3

Machine the mortises and clear them out with a 6mm chisel ensuring that they are clear of chips. Now cut out the waste material from the half laps either by hand or by setting up the mitre saw with a depth stop (see Chapter 3 Key Construction Skills for more information).

With the half laps cut, test fit the glazing-bar assembly, making sure that all the joints fit nicely, and make any adjustments with a chisel until the shoulders are flush. Dismantle the assembly and apply glue to the joints ensuring that it is spread evenly over the surfaces, then refit the bars together. Use a speed clamp on each joint to hold it tight while the glue goes off, ideally overnight.

If you want to secure the joints further to enable you to work on the frame before the glue is dry, you can use a carefully placed screw on the back; however, I prefer to use a single star dowel, as there is no risk of splitting the timber in such a small area.

STEP 4

The next step is to cut down the tenons on the top and bottom rail to fit the stub corners, so mark 20mm off the outside edge of the tenons and cut them to size with a tenon saw. Make sure that the shoulder is perfectly flat and square where the tenon is cut away.

The frame can now be carefully assembled by first attaching the top and bottom rails to the glazing bar assembly, followed by the stiles; apply

Step 3.

Step 4a: The tenon has been cut down.

glue to all the tenons prior to doing this. Use sash cramps to close up any gaps and square up the frame, but put the frame face down in the clamps so that the joints can be fixed from the back. Now work your way around the frame using star dowels to secure each joint. Put two in a staggered pattern on each corner, remembering that it is a stub tenon, so avoid the area nearest the corner.

Step 5

Remove the frame from the cramps, and sand any uneven joints flush. Use a block plane to arris the outside edges of the frame. We are now going to use the ovolo cutter in the router to add the moulding detail to the glazing bars, so set up the depth of the cutter on an offcut until the finish is to your liking. The most challenging aspect of this is keeping the router balanced while you work your way around the glazing bars.

Trying to hold the router perfectly level on a 20mm-wide batten is not easy, so the best solution is to use some spare 28mm timber as a support block, and screw it to the bench. This will

Step 4b: A sash clamp is used to close up the joints, then the star dowels are fixed in.

Step 5.

keep the router base nice and flat while you work your way around the edges of the frame. As we are working around the inside of each pane, the router should be used in a clockwise direction to ensure an even finish. Take your time with this process, and go back over any areas that have not taken the cutter properly.

Step 6

As you will notice, the corners of each intersection have the tell-tale 'I've been routed' look, in that they are rounded, which spoils the overall effect. To fix this, some careful work with a chisel is required. On each corner, and using only hand pressure, cut a line across the grain first into the corner with the back of the chisel in line with the detail, and then along the grain. Now turn the chisel bevel side down and very gently lift out the corner working with the grain. Extreme care is needed, particularly on the glazing-bar intersections, otherwise you will chip pieces off very easily!

A similar effect can be achieved where the quadrant section of the moulding meets. This can be painstaking work, and might not be necessary in all situations, but it makes a huge difference to the overall look of the window.

The frame can now be finished up, so arris any missing areas and clean up any fluffy edges from the routing process. The frame is ready for glazing.

Step 7

In this example I am simply screwing the polycarbonate to the back of the frame. Cut it to size, and double check the dimensions a couple of times, as it is expensive stuff! Leave the protective film on the polycarbonate for as long as possible, as it will save any wear and tear while it is moved around the workshop. Lay the frame on top of the polycarbonate and draw around it with a whiteboard marker, being careful not to let the ink soak into the timber. Put the frame aside and drill countersunk pilot holes into the plastic around the

Step 6.

Step 7.

The polycarbonate is trapped between two frames.

A slot cutter makes it easy to fit plastic or panels into the timber frame.

outside edge set centrally within the frame width, then screw it to the back. Don't overtighten the screws: set the torque collar on the drill so that the screws just pull tight enough without distorting the plastic. Unscrew the plastic, remove the film, and refit once on stage.

There are other ways to fit the plastic to the frame, depending on the design: sandwiching the plastic between two thinner frames is a good way to do this. If the edge of the window is seen by the audience, then one side of the frame can be rebated and the plastic set in. This is really useful for French windows and doors.

It is also possible to machine a slot around the centre of each part of the edge of the window frame and trap the plastic when the frame is assembled; this is more practical for windows with open panes than this style – although still possible, it just requires some careful marking out and planning prior to construction.

Another useful method is to trap individual panes in the centre of the frame using a small quadrant or angle, similar to fitting panels into a door. This can also be quite laborious on windows with multiple panes like this, but looks absolutely fantastic under lights, as each pane will have a slightly different reflection and look more like a real window than using a solid piece of polycarbonate.

Building a Window into a Flat

Building windows into flats is a very similar process to adding a door frame, so have a look over that section first. The window will need reveals to give the impression that the wall is thicker than it actually is, so a reveal frame should be constructed and fitted into the flat frame. The main difference with window reveals is that they are not sitting on the floor, so will likely require additional support underneath to hold up the window. This

A curved window reveal.

is particularly important if the window is practical, as actors climbing on to the reveal or sitting on the window sill will easily damage the flat if this has not been considered.

Interior windows should be fitted with a window-sill, which could be as simple as adding a length of half round to the edge of a piece of plywood. For a realistic look this should be extended past the width of the aperture slightly, and the ends rounded to match the profile.

Many windows (and doorways for that matter) feature curved reveals. These can be constructed using plywood formers with spacing battens, and clad with flexi ply. Have a look at the column project (Chapter 9, Project 10 'Build a Column') for an idea of how to approach this. As with doorways, it is generally best to fit the reveals into the flat prior to cladding, particularly with windows, as they won't usually feature moulding around the outside of the aperture to hide exposed edges.

8
STAGE FLOORS, TRUCKS AND PLATFORMS

A typical method of moving scenery onstage is to mount it on a movable base known as a truck. These can be constructed in many different ways depending on the type of movement required, the method of movement (automated or manual), and the size and weight of the scenic piece. We are going to look at a reliable method for constructing a basic timber-framed truck base. The base is constructed from 6 × 1.25in (145 × 28mm) timber, and fitted with an 18mm plywood top. Below is a plan view and side elevation of the truck: notice there is additional hardware fitted in the form of swivel castors. It is important to allow enough clearance between the truck frame and the stage to allow for cables, ramp access and manual handling. In this case the clearance is 18mm, but this should be adjusted to suit the particular requirements of the production.

Project 8: Build a Timber Truck Base

You will need the following:

- Combination square and pencil
- Tape measure
- 18mm plywood
- 145 × 28mm (6 × 1.25in PAR) timber, ripped down to 125mm
- Chalk line

OPPOSITE: The prison truck from *Measure for Measure*, RADA Jerwood Vanbrugh Theatre. Director: Jonathan Miller; designer: Lorna Ritchie; photographer: Mark Tweed. © RADA

- Straight-edge
- Handsaw
- Cordless drill
- Countersink drill bit
- PZ2 driver bit
- PVA glue
- Brad gun
- 5 × 60mm screws
- Block plane
- 120g glass paper and block

Method

STEP 1

We are going to construct the truck using the 'footprint' method, which means that the first stage is to mark out the plan view of the truck frame, full size, on to the 18mm plywood sheet in order to give us a template from which to work. To ensure consistent measurements, I will use both the short and the long 'factory edge' of the boards as the datum points.

STEP 2

Mark out the corners of the frame, and on the top edge use a straight-edge to indicate its position. You may want to draw in a line with a Sharpie to make it slightly more permanent, particularly if you will be walking on the sheets a lot during construction. Next, mark the centres of the internal rail of the frame, working along each long edge, and then connect the points.

any small discrepancies as you go. With the frame assembled, work your way around screwing each joint tight.

Step 8

With the truck frame assembled, the next stage is to fit the top. Much like cladding a flat, use the factory edges of the board to square up the truck frame. The added benefit of working from a footprint is that you can now use the mark-out as a guide for where to put screws to hold the top down.

First trim the boards using a circular saw or panel saw. With the top cut to size, apply a bead of glue to the top of the truck frame and lay on the first board, ensuring that the short edge of the sheet is flush with the frame. Now clamp it in place, and work your way along with your countersink drill and driver, fixing the top down with 5 × 30mm screws equally spaced. Now, working along the long edge of the board, ensure that the edge remains flush as the screws are fixed in.

With the two factory edges secured, the rest of the screws can be fixed. Be careful to space them evenly, and keep screws away from any butt joints to avoid clashing with other fixings. The truck is now ready for the hardware to be fitted.

CASTORS AND BRAKES

There are several considerations to make when deciding the type of castors to use on a truck or revolve. Look out for the following: the weight rating, the wheel diameter and overall height, whether swivel or fixed, and the wheel type.

The Weight Rating

The full weight of the scenic element and the truck itself should be calculated, as well as taking into account what the maximum load on a single point may be.

Step 7.

Step 8.

Example 1: On a scenic piece with the weight evenly distributed over the truck fitted with four castors, dividing the total weight by three would give a good indication of what the rating of each castor should be, also building in a safety factor. So if the total weight of truck and scenery is 600kg (1,320lb), dividing by three tells us that the castors should have a minimum load rating of 200kg (440lb) each.

Example 2: On a truck intended to carry people, or where the weight of the scenery is not evenly distributed, it is important to select castors capable of supporting the maximum possible load each. For example, if our truck was intended to carry four actors weighing a total of 320kg (700lb), we should assume that their weight will not always be evenly distributed over the truck as they could easily all stand on one corner or move around, in which case each castor must be capable of supporting the full weight of the actors plus the weight of the truck. Therefore, if the truck weighs 20kg (44lb) and the actors 320kg (700lb), then the absolute minimum rating for each castor would be 340kg (750lb). I have selected castors with a rating of 350kg (770lb) each, which would be suitable for both examples explained above.

Again, a proactive approach to risk assessment is crucial here; the examples above are hypothetical, and whilst a good guideline, cannot account for every possible situation. Thorough consultation with the production team, and reference to ABTT guidelines, is important to ensure safe working practice. The Flints catalogue has thorough details of all their products, especially castors, so I would highly recommend paying their website a visit.

Wheel Diameter and Overall Height

The larger the wheel diameter, the easier it will be to roll the castor over uneven surfaces and the smoother the move will be, although turning may be more of a challenge than on a smaller wheel. The benefit of a smaller wheel diameter is that the castor will generally have a lower profile and will therefore require a less bulky truck base to hide it, as well as being easier to manoeuvre. The example I have chosen has a wheel diameter of 100mm, and the overall height of the castor is 125mm.

Wheel diameter and overall height are closely linked: obviously the larger the wheel, the taller the castor will be overall, so make sure that the truck frame will fit the castor effectively. It is typical to leave a small gap between the stage floor and the underside of the truck base to give enough clearance for uneven surfaces, cables, fingers and suchlike.

As seen in the section view through the timber truck with the castor fitted, notice how there is 18mm clearance underneath, and the castor has been blocked up and mounted to make up the difference between the overall height (OH) of the castor and the height of the truck.

Swivel or Fixed?

One of the most important considerations when selecting castors is thinking about the movement required of the truck. Swivel castors offer a degree of flexibility and freedom of movement as the centre of the wheel is offset from the centre of the mounting plate, which drags the castors into position when the truck is moved. Whilst this offers a lot of freedom of movement, inevitably the truck will 'squirm' around as the castors align, meaning that a truck fitted purely with swivels is not ideal for linear movement or in restricted space. The castors require extra space around them within the frame to rotate, and because of the variable positions into which the castors can move independently, precise repetitive actions can be difficult to achieve without a carefully choreographed set of movements.

Fixed castors, on the other hand, are perfect for linear or rotational movements, such as a stage truck running on a track or on a revolve, where precise predictable movement is required. Accurate fitting is crucial with fixed castors, as they

STEP 3

Carefully drill a pilot hole in the centre of each bolt position through to the top of the truck. Now switch to the 25mm flat bit and drill a shallow hole from the top of the truck, using the pilot hole as a guide for the spike on the flat bit. The purpose of this is to make a small recess for the domed head of the coach bolt to sit into, to help it sit flush and prevent it from crushing the ply when it is tightened – so there is no need to go very deep, just 3–4mm should do it.

STEP 4

Now the main bolt hole can be drilled out using the 11mm drill bit; using a marginally larger size than the 10mm bolt makes it easier to push the bolts through. Now push the coach bolts through from the top of the truck, and put the castor in position. Hand tighten it up, having put a washer and nut on to each bolt. Now strike the domed head of each bolt with a hammer so that the square collar is seated into the ply. Try to do this with only one or two strikes, as repeated hits will create a rounded hole for the collar and the coach bolt will not tighten properly, which can be very annoying! Hand tighten the nuts again to take up the slack.

STEP 5

Finally position the castor ensuring that it is seated squarely, and tighten up each nut with the ratchet, alternating diagonally to ensure even pressure. Keep tightening until the domed head of the coach bolt has sunk to the desired depth, but be careful not to overdo it or the thread of the bolt may interfere with the rotation of the castor. If this happens, simply cut back the excess thread with a hacksaw or angle grinder, and the job is complete. ·

ROSTRA

Rostra are essentially structural frames put on stage in order to change the height or shape of the stage itself, and there are many different styles. Most theatres and scenery companies

Step 3.

Step 4.

Step 5.

Most theatres have a stock of modular decking, which offers versatile layout options on stage.

use modular steel rostra, available to hire or purchase from companies such as Steeldeck or Scott Fleary, for much of this work. These are very versatile and hard wearing (though very heavy compared with timber!), and are simple to work with due to the fact that they can be fitted with standard-diameter scaffold tubing (48.3mm OD) legs, and are easily bolted together. As standard they come in 8 × 4ft sizes and variations thereof, such as 4 × 4ft, as well as custom shapes and sizes for a variety of applications, and you will no doubt encounter them on many jobs across the industry. They are the first choice for heavy-duty staging applications.

Traditionally timber rostra were used extensively, and still are, although the design has evolved somewhat due to the rise of CNC machining. Years ago most theatres would have a stock of 'folding gate rostra', which are made up of mortise-and-tenon frames, hinged together so they can fold flat. They are then fitted with a plywood top that can be blocked in position, and multiple rostra could easily be bolted together. (*See* John Blurton's excellent book *Scenery: Draughting and Construction* (Routledge, 2001) for detailed information on how these are made.)

However, as fewer scenery shops use mortise and tenon as their framing method, this has become less common, and a variation has emerged in the form of plywood rostra. They are an evolution of the same idea, but utilize plywood gates instead of timber frames, which can easily be cut out with the CNC router or power tools, and assembled by hand. Although the material cost is higher, the time they take to fabricate is considerably less, so they have become the industry standard in carpentry terms.

In the following project I demonstrate how to make a small raked plywood rostrum, a guide that can be adapted to suit your own project, and can be used in conjunction with steel rostra and timber platforms, infills and treads, to achieve a huge variety of shapes and configurations. I have made this project more challenging by making it triangular as well as raking, meaning that there are some different angles to think about. Be sure to try out some of the earlier projects first before taking on this one.

Project 9: Build a Raked Rostrum

You will need the following:

- 18mm plywood
- Adjustable bevel
- Combination square
- Tape measure
- Pencil
- Brad gun
- Drill with countersink pilot bit
- 4 × 45mm screws
- 5 × 30mm screws
- PVA glue
- Circular saw or table saw
- Jigsaw
- ¼in router with chamfer bit
Roofing square

Project 9: Build a Raked Rostrum.

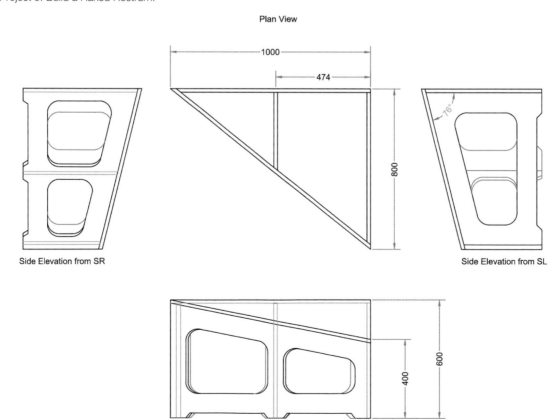

Plan View

Side Elevation from SR

Side Elevation from SL

Front View

Method

STEP 1

Study the technical drawing, draw out a footprint on a sheet of plywood, and write a cutting list of the component parts that make up the frame of the rostra. Leave the top out of this for now – we will look at that later. We will cut these pieces slightly oversize initially, and then cut in the angled tops on the gates and the bevels on the frames afterwards, so allow an extra 15–20mm to the height of each piece to accommodate this.

Think carefully about how to make the best use of the sheet material you have available to avoid unnecessary waste, and cut the boards to size. Be sure to label the boards with a letter, and mark the top with an arrow to avoid confusion later on.

STEP 2

The next step is to mark out and cut the angled end cuts on to the gates. This dimension is given on the drawing here, but it is better to use the footprint to mark out the angles and set up your bevel.

Pro Tip: Once you have set your bevel, mark your reference angle on to an offcut so that, should you accidentally drop your bevel (something I do regularly), you can quickly reset it to the correct angle.

Measure up from the bottom corner on each side of the gates, mark the heights, and then connect the points with a straight-edge. Now cut in the angles. Some of the end angles are too acute to use machines or power tools, so cut by hand and tidy up with a plane.

STEP 3

With the ends cut accurately, the angled top of side gate B can now be marked in and cut with a circular saw. This will allow you to then set the bevel accurately in order to set up the circular saw for cutting the compound angle into the top of gate C. Place the bevel perpendicular to the end

Step 1.

Step 2.

cut on gate B to find the correct angle, set up the saw, and cut gate C to the correct dimensions.

The bevel can now be cut along the top of the back gate A. We know from the drawing that the angle is 6 degrees, but if the angle were not provided on the drawing you were working from, then you would solve the problem by simply setting your adjustable bevel to the top of gate B in order to use it as a guide for your table saw or circular saw.

STEP 4

Now that all the pieces have been correctly dimensioned, work can begin cutting in the detail. In similar style to the block treads project in Chapter 6, we are going to remove large sections of plywood from the centre of each gate to save weight and improve access inside (see feature box 'Cutting the Guts Out', also in Chapter 6), as well as cut feet into each piece. There is no need to have a solid slab of plywood (apart from the rostra top, obviously!) as long as ample width is left on to provide a 'frame' shape in each piece. The rostra will sit better on feet rather than solid edges, particularly on uneven surfaces.

Draw a radius on each corner of each void for a nicer finish, in this case with the trammels set to 60mm. Carefully remove the waste with a jigsaw, and tidy up if necessary. Finally run the chamfer cutter in the laminate trimmer around any exposed edges to obtain a neat finish – though be careful of the bevelled edges, which are vulnerable to damage.

STEP 5

Begin assembling the rostra by attaching the side gate (B) to the back gate (A). Apply a thin line of glue to the edge of the gate and offer it up, ensuring that it is perfectly flush at the top, as well as lined up to the pencil marks on the footprint. Starting at the top and working down, brad the gate into position, adjusting as needed as you go. Repeat the process for the centre gate (D), again ensuring that it matches the footprint and lines up with the bevel on the back gate (A).

Step 3a: Set the bevel against the cut on the side gate.

Step 3b: The compound angle is cut along the top edge of the front gate.

Step 4.

Step 5.

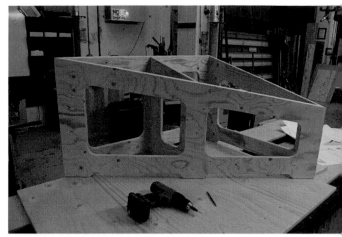

Step 6.

Step 7.

Step 8.

STEP 6

The front gate (C) can now be attached, ensuring that the bevelled ends line up nicely and the frame matches the mark-out. Now work your way around the frame with a countersink drill and make pilot holes for the screws over the whole piece, and screw it together. Don't get too carried away – three or four screws per join is plenty.

STEP 7

Look down the rake at the top of the frame and check that all the joints are flush. Any discrepancies can be planed out at this point prior to fitting the top.

STEP 8

The top can now be cut to size. Notice how the actual width of the top needs to be wider than the width of the rostra in plan view, due to the fact that it is sitting at an angle. It will also need to have bevels cut into the downstage and upstage edges to match the rake, so over-cut the width to allow for this. The dimensions for this can be found on the drawing, as the bevel angle will be the same 6 degrees we were working with earlier. Mark out the cuts and use either a circular saw or table saw to trim the edges.

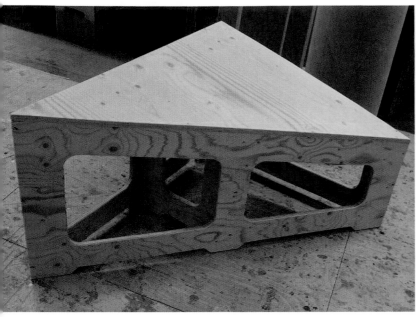

The finished rostra – front view.

The finished rostra – side view.

Here, Laura is fitting underfelt to rostra tops in the National Theatre workshop.

STEP 9

The frame can now be glued, and the top can be screwed on. Watch your screw spacing, and be sensitive to the fact that your screws should be driven in perpendicular to the floor rather than the top, which will prevent screws poking out of the side by accident due to the angle of the rake. Once fixed, the whole rostrum can be sanded, and any protruding edges planed in. The top should be arrised by hand in this case, as the bevelled edges will affect the router's ability to shape the edge neatly.

Pro Tip: Underfelt is often applied to the underside of rostra tops to deaden the hollow drumming sound caused by footsteps and vibrations on stage. If required, this should be applied to the whole board prior to blocking the top on, if the top is removable. If the tops are permanently attached, then the felt can be cut to fit the open sections on the underside. It is simply stapled down using a wide crown stapler, and trimmed with a Stanley knife – be sure to buy the fire-retardant type!

INFILLS, DECKS AND PLATFORMS

A common way of filling in smaller spaces between decks is to build small boxes such as the plywood rostra; however, sometimes there may be gaps of less than 300mm wide to fill, where making a whole box is overkill, or where the decks are too high off the stage to use this method. There are a couple of simple solutions to these situations: to infill the gap with a plywood insert, or to make a timber 'deck'.

Infills

By bolting timber battens to the edges of the corresponding steel rostra, set 18mm below the top of the decks, a plywood insert can be cut in and screwed down to the battens without the need for

A basic ply insert supported by battens bolted to the edge of the deck.

legs. This is a good way to deal with tricky gaps, particularly if they aren't perfectly square, as the insert can be planed in for a very neat fit. It is really important that all the edges of these inserts are supported, because otherwise they may become a trip hazard, or worse, fail completely.

Timber Decks

Another solution for staging is simply to make a timber 'deck'. This is particularly useful on the off-stage edges of the set, or on areas with less regular shapes where a plywood insert wouldn't be supported on all sides, or where the gap between decks is larger than 300mm and the top of an infill will require extra rails for support.

The method is very similar to the truck base project in Chapter 8, so refer to that project for more detailed information on construction. Material choice for the frame could be scaled down depending on the size of the infill and the depth of the corresponding deck edges; the important

Simple timber 'L' legs are perfect for lower infills and platforms.

feature is to frame the plywood top so that it is nicely solid to walk on and can be securely bolted to its neighbouring decks.

Any unsupported edges can be made good by adding in simple timber legs. Single battens are not recommended as they can flex quite easily. The best method for making reliable timber legs is to use two pieces of 3 × 1in, butted and screwed together in an 'L' shape. These can then be bolted through the frame of the deck, as well as fixed to the stage.

Platforms

As useful as steel rostra are, it is not always practical to use them as a free-standing platform without extra bracing on the legs, making them cumbersome to move in a scene change, for

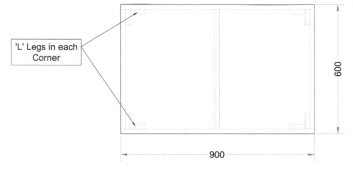

'L' Legs in each Corner

600

900

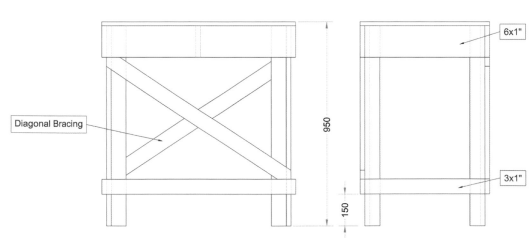

Diagonal Bracing

950

150

6x1"

3x1"

A simple timber platform with braced legs.

example, due to their weight. You are also limited to the range of sizes in stock, so for some situations making a timber platform is the better choice.

Essentially it is built in the same way as the timber deck, fitted with 'L' legs as described above. However, as shown in the drawing, it also requires some extra support to brace the structure and keep it stable, and the legs must be tied together to prevent them from slipping apart. Notice the diagonal braces in place to prevent the frame from twisting, and the leg ties near the feet.

Pay careful attention to the height of the platform relative to the size of its footprint when making pieces like this. No matter how solid you make the structure, a freestanding platform which is taller than it is wide is at risk of falling over without additional bracing or being securely fixed to the stage, so be sure to plan your builds with this in mind.

Ramps

Ramps can be constructed in a similar fashion to plywood rostra, but pay particular attention to what the ramp is being used for, and tweak the layout of the formers to suit. Access ramps may be subjected to quite high point loads from castors on trucks, for example, so be sure to add enough support with this in mind, in some cases increasing the thickness of the top to 25mm or using stronger plywood such as birch to reduce the chance of punching through it.

One of the more challenging parts of constructing a ramp is cutting the acute angle where the end of the ramp meets the floor. A good way to achieve this with hand tools is to plane the angle into the edge. Lay the ramp top flat on the bench, upside down, and mark the position of the back of the angle with a square. Try to line up the end of the board with the end of the bench, and clamp it down. Begin planing

Basic stage ramp.

Plan View

Front View

Side View

A stack of birch plywood show floorboards at the Royal Opera House workshop. The edges are slotted to allow metal flushing plates to locate the sections together.

A CNC-cut show floor laid out in the National Theatre paint shop.

Consideration should be given to any surface detail as well; many floors will have CNC detail routed in, so a suitable thickness with this in mind should be chosen. As an absolute minimum 6mm could be used, but 12mm or 18mm would be the best option if budget will allow.

Tips for Laying a Show Floor

There are many approaches to this, many of them situational, but there are a few guidelines to follow that will help the process along:

Step 1: Check that the boards are square! Factory boards are not always perfectly square, so expecting fifty of them to tessellate together perfectly without measuring them up first is asking for trouble. The easiest way to do this is to measure the diagonals of the board: if the measurement is the same, the board is square, if not, then machine the boards square and adjust the floor layout to the new dimensions.

Step 2: Break it down sensibly! With heavily patterned flooring such as parquet or planks, break up the floor up into larger sections to avoid laying individual planks in situ, for example, but make sure that the break lines follow the style of the floor. Seeing 8 × 4ft rectangles cutting through a beautiful mosaic will ruin the illusion, so plan the breaks carefully. CNC is fantastic for show floors, and often worth spending the money on for complicated patterns.

Step 3: Protect the sub floor! Productions in some venues may require the playing space to extend on to areas not normally used as a stage. It is really important to protect the floor underneath from damage from nails or screws as well as paint

Sam and Courtney laying out a CNC'd MDF parquet floor on stage in RADA's Jerwood Vanbrugh Theatre.

washes seeping through gaps, so factor this in to your fit-up time and budget.

Step 4: Start in the middle! The ideal place to begin laying the floor is the centre line on the down-stage edge where it crosses the setting line. Working out from centre stage back and out into the wings will mean that if discrepancies develop they can often be rectified more easily. If there is no clear centre or setting line, then mark in your own reference lines with a chalk line using the '3-4-5' method found in Chapter 3. Always try to lay the boards in a staggered pattern to offset join lines: this will go a long way toward making the floor appear seamless.

Step 5: Don't go crazy with the fixings! Use a sensible pattern of fixings to hold the floor down, and only if you need to fix it at all. Avoid getting too 'trigger happy' with a nail gun as this will mean a painful get-out trying to remove them all again. If the show floor is being painted on site, then expect it to move slightly, and consider allowing tolerance between sheets by using skin ply spacers. The boards will swell and contract when they get wet and dry out again, so fix them with this in mind – don't just fix them around the edge, and try to go for a consistent pattern on each board. Be sensitive to the design, and if possible use fixings to match the pattern.

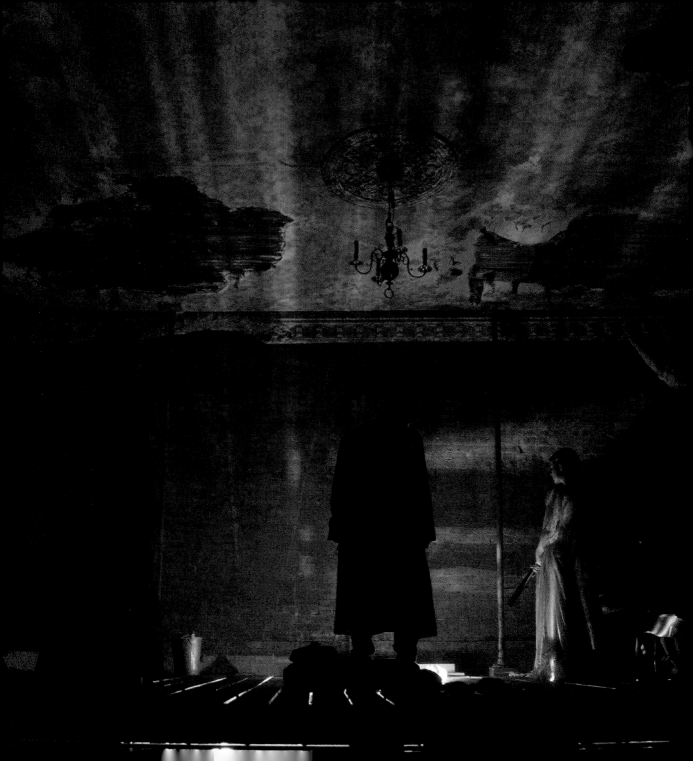

9
ARCHITECTURAL FEATURES

MOULDING

There is a wide range of styles of decorative moulding, which can be a really effective way to give your sets more depth and realism. It is possible to make your own if you have access to a router table or spindle moulder, but common patterns such as torus and ogee are readily available from most suppliers.

OPPOSITE: *Bloody Poetry*, GBS Theatre. Director: Sebastian Harcombe; designer: Sarah-Jane Prentice; photographer: Linda Carter. © RADA

Build up a stock of different moulding patterns to suit a wide range of period styles.

BUILDING A LARGE CORNICE

Cornices are a common feature on period scenery, traditionally fitted between the join of the top of a wall and a ceiling, and often very decorative; they can give a fantastic sense of depth and grandeur to a set. In reality these may be carved from solid wood or stone, or moulded in plaster; however, none of these options is really suitable for use on stage, so a few different approaches have been developed over the years to replicate them and make them practical to build.

The use of vac form is one option, and companies such as the excellent Peter Evans Studios specialize in providing designs from different periods. Their vac forms are available in a range of different plastics and thicknesses to suit most budgets. Vac form can be difficult to work with at first, however, as it generally requires support to keep it looking square and consistent – for example, mitring corners neatly can be very challenging – but in certain applications it is a quick and reliable way to add detail to your set, and is definitely worth considering.

Quite often, however, a bespoke design of cornice is required, perhaps borrowing elements from more than one period or style, where materials may need to be combined to achieve the effect required by the designer. The best way to approach this kind of project is to make up a sub-frame using a series of formers, spaced apart with battens. The detail of the design can then be applied in layers, stacked up to form a much deeper shape than would otherwise be achievable. This way the weight of the piece can be managed as it is predominantly hollow, and there is plenty of scope for adding fixing

This huge opera cornice is made up of hard-clad plywood formers, with layers of polystyrene and timber moulding built up to give it depth.

points or flying hardware to the back. The principle is the same, regardless of the design, so try out different combinations of mouldings and materials to achieve different effects.

The first step is to create the master template of the formers. The first stage of this is to draw a side view of the profile of the finished cornice, by marking it out full size on a piece of 18mm plywood.

Now think about how you might layer up the detail on the face using different thicknesses of timber and mouldings, and begin to mark these out on to your board, hatching them out to leave the remaining shape of the former. Cutting small slivers of moulding you have in stock can be really useful here, to help lay out the shapes and try out different combinations. There is a large radius to cut in this case, which is too big to use moulding, so flexi ply will be added to the formers. I have added 5mm to the radius to account for the cladding. Once the detail has been marked out, plan where the supporting battens, which will run through the formers, will sit. In this case we are using 2 × 1in on edge, which is placed in the top corner and bottom edge, so the profile of this is also marked on the former and hatched out.

With the master former marked out, the next stage is to neatly cut it out. We will use it as a template to trim the remaining formers to, so it needs to be as accurate as possible. I used a combination of the crosscut saw, the mitre saw and a jigsaw to cut out this one, and gave it a careful sand in places where needed.

Blanks of 18mm plywood are cut for the remaining formers, slightly larger than the actual size required, and the template traced around on to each one. These can now be cut out leaving the outline on the ply, which will provide just enough material for the router to trim neatly without struggling. A laminate trimmer or ¼in router is probably best for trimming the formers, as it will allow more intricate cutting. Tape or screw each former to the template and trim them to size, then use a chisel to square out any round corners where the router is unable to cut.

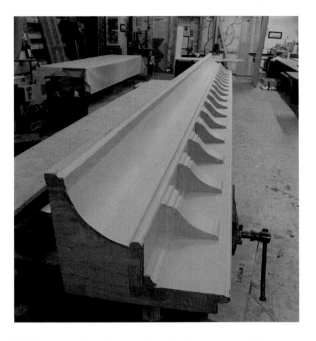

This large section of cornice, built by James Russell, features dentils each with scotia moulding around them, as well as several layers of built-up detail. Time consuming to build, but the finished piece is very impressive.

This cornice sample, built by Steven Peters, demonstrates how many different material thicknesses and shapes can be utilized to create different forms. All the moulding on this piece was made using a router and standard timber stock. Notice that a void has been cut into the former to save weight.

Janine test fitting a mitre joint prior to fixing.

The battens are then marked out with the placement of the formers. The mark-out can be transferred over to each length of moulding, which will help to fix it accurately and ensure that everything is square. With the positions marked on the battens, they are pre-drilled and countersunk, and the formers attached with screws and glue.

Now the final details can be added to the face of the cornice. In the case of the large cornice above, dentils were added at equal intervals, so a spacer template was cut and the dentils fixed in position from the back. This is the stage to add smaller details such as vac form sections, or cast features perhaps. The whole piece can now be carefully filled and sanded, ensuring that the mouldings and mitres are kept looking sharp and are not overly softened by overworking them.

The cornice section is now complete and ready for use. If a mitred join is required, it is advisable to make an angled former to act as an end plate. This will help to tie everything together and pull the joint tight where it meets the other half. An excellent drawing showing how to calculate and draw out a cornice end plate can be found in *Scenery: Draughting and Construction* by John Blurton (Routledge, 2001).

Project 10: Build a Column

This simple project demonstrates a reliable way of creating curved surfaces using timber and sheet materials. In it I am building a simple circular column, the basic principles of which can be incorporated into many different types of work.

You will need the following:

• Tape measure
• Pencil
• Trammels
• 18mm plywood
• 22 × 22mm PAR timber
• 5mm flexible plywood (long grain)
• Glue
• 4 × 40mm screws
• Brad gun
• Narrow crown stapler
• Chalk line
• Laminate trimmer with a flush trimming bit

As you will see from the drawing, the column is constructed by spacing apart a series of formers using timber battens set at equal intervals around

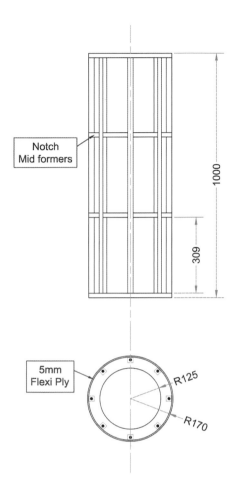

Notch
Mid formers

1000

309

5mm
Flexi Ply

R125

R170

Project 10: Build a Column.

Step 1: The former is marked out and divided into eight.

the circumference. These are set in so that a thin flexible plywood, in this case 5mm thick, can be used to clad the cylinder, producing what appears to be a solid column. This project is a good exercise in making multiples of the same component, as well as learning to clad curved surfaces.

Method

STEP 1

The first step is to produce one of the formers. We will use this as a template to make the rest of them, so take your time to ensure this one is as accurate as possible. Start by marking out a

circle using the trammels set to the outer radius of the former, in this case 170mm. Next we want to mark the internal radius of the former, which is 45mm smaller. At this point it is a good idea to divide the former into eight, which will provide the intended location of the battens. For a step-by-step guide to this process, refer to the 'Useful Geometry' section in Chapter 2. If you skip this part, it becomes difficult to mark out these positions accurately, especially when the middle of the former is cut out and the centre point has gone!

The number of divisions and therefore support battens may vary depending on the diameter of the piece you are constructing. The flexible plywood requires support at regular intervals – certainly leave no spaces larger than 400mm, otherwise the cladding will be vulnerable to damage and may not conform evenly around the formers.

With the circle divided up equally, the dividing lines should be offset on either side to represent the width of the battens. Here I am using 22mm timber, so I have offset each line by 11mm on each side.

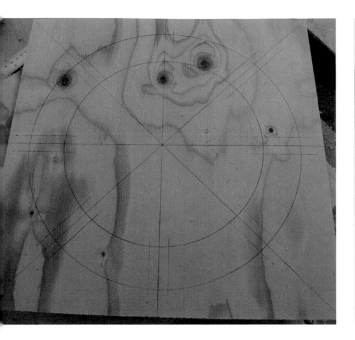

Step 1: The batten widths are marked on to the former.

CNC routers are fantastic for making multiples of intricate shapes, such as these formers.

STEP 2

The former is ready to be cut. This can be done using a jigsaw or perhaps on the bandsaw using a circle jig (if the radius is small enough to fit through the throat of the bandsaw), but in my opinion the method that will achieve the neatest and most accurate result is to cut out the former using a router fitted to a trammel arm or radius jig. The benefit of this method is a reliably clean finish, and the fact that it can be scaled up to almost any size you are likely to need. The exact design of the trammel arm will depend on the router you are using, there are many power-tool brands that sell this type of accessory for their kit – but they can be expensive.

MAKE A SIMPLE ARM FOR SMALLER WORK

For smaller work and repeat cuts of the same radius, a simple arm made from plywood can be directly bolted to the base. It has a hole cut out large enough for the router cutter to sit through with some clearance, and the pivot point is carefully measured from the outside of the cutter. The downside of this type is that it is limited to shorter radii. For larger arcs, a metal trammel guide is recommended, which can be easily extended to larger dimensions.

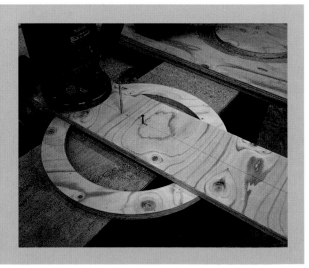

A simple trammel arm for the router made from plywood.

Due to the fact that we are cutting out a full circle, it is important to cut the outer diameter first, to retain the centre point for the second cut. A standard fluted cutter is required for this, the diameter of which will depend on the material and radius being cut. For example, ploughing through a sheet of 18mm birch plywood with a 6mm cutter will almost certainly lead to it overheating or breaking.

Screw the pivot point of the trammel arm down to the centre point, and ensure that the inside edge of the router cutter is sitting perfectly on the pencil line. Due to the fact that the plywood is 18mm thick, we will cut out the former in a series of passes so as not to put too much pressure on the cutter, which will prevent it from overheating. Here I will go around the circle in three passes, in 6mm increments each time. It is a good idea to pre-set these increments using the stops on the base of the router if possible.

Plunge the cutter into the board to the first depth stop, and make the first full pass around the former. Next clear out any waste and dust from the channel you have created, and repeat the process until the circle has been cut.

STEP 3

Repeat the process for the internal diameter. This time, both the former itself and the central offcut need to be fixed down securely to avoid accidentally damaging the workpiece as it becomes free. The radius should be set from the pivot point to the outside of the cutter this time.

As you progress through the cut, care should be taken to ensure that both pieces are still stable. If needs be, a small 'bridge' can be left on between the two parts and then cut off by hand.

With the former cut out, give it a light clean up with some sandpaper if needed, and then transfer the eight batten positions on to the edge of the piece.

STEP 4

The next stage is to mark out and cut the remaining formers. They can be marked out and cut in exactly the same way as the previous step if desired, but I am going to explain a method using

Step 2.

Step 3a: The centre is cut out next.

Step 3b: Then the batten positions are transferred on to the edge of the former.

Step 4.

Step 5a: Make sure the formers are firmly clamped together before trimming.

the flush trimmer fitted to the router to copy the original former which is our template, another useful process for a variety of projects not limited to cutting circles. This is a better option when dealing with more than a handful of pieces.

Draw around the template on both edges on to a board three times, spaced as tightly as is practical in the interests of being economical. I usually leave a 5–10mm gap between the pieces, or away from the edges of the board.

The formers can now be roughly cut out with a jigsaw, leaving a couple of millimetres extra around the edges to allow for easy alignment. This will be trimmed flush against the template in the next stage, so accuracy isn't so important, as long as you don't cut over the lines!

STEP 5

Take each former in turn, centralize it on the template, and secure it either with screws or nails, or double-sided tape. Clamps can now be used to secure your work to the bench, making sure to have the template on the bottom so that the bearing of the cutter will run around it.

Unfortunately, particularly on smaller pieces, the clamps will be in the way a lot of the time, so just work around the piece a section at a time, moving the clamps as you go. Another solution to this is to cut a series of timber blocks to lock the workpiece in place, essentially building a jig on the bench, which is a good time saver if you have many pieces to work through.

When trimming the inside of the piece, go clockwise, and vice versa for the outside – this will give you the cleanest finish, as the direction of travel will oppose the direction in which the router cutter is spinning.

With all the formers cut, transfer the batten mark-out from the template on to them. Put two of them aside – these will be the top and bottom formers, and they require no further work before assembly. The remaining pair needs to be notched out to allow the battens to sit flush with the outer perimeter, so using a combination square, mark in 22mm on each position, and then carefully remove this area with a jigsaw. Use a small offcut of the batten material to check the fit as you go.

Step 6

The battens themselves can now be cut to length, which will be the total length of the column, less the thickness of the top and bottom formers. In this case the calculation would simply be 1,000mm – (2 × 18mm) = 964mm. So on your cutting list express this as:

22 × 22mm PAR
8 @ 964mm

Now clamp the battens together, ensuring that they are flush with each other, and mark out the positions of where the formers will sit, as indicated on the drawing at the start of this project.

Pre-drill single pilot holes into each position, as well as around the top and bottom formers where the battens will sit. The column is now ready for assembly.

Step 7

We want to begin by attaching the battens around the pair of central formers, and by working on opposing sides. So begin by gluing and bradding the first batten into the top positions (12 o'clock) of the formers. Ensure that the marks line up perfectly each time. Rotate the pieces around and apply the second batten in the opposing notches (6 o'clock). Repeat this process until all eight battens are fitted.

Now the top and bottom formers can be attached following a similar approach. With the frame assembled, work around it making any necessary adjustments to alignment, before putting in screws to tighten up all the joints. Wipe off any excess glue and tidy up any edges that aren't perfectly flush with a plane or sander.

Step 5b: The notches have been cut for the battens to sit into.

Step 6.

Step 7.

The frame is complete and ready to clad.

STEP 8

Now for the fun bit: cladding! Our column is 1,000mm tall with a diameter of 350mm. We want to allow a little excess material in both directions when cutting our sheet, so I will cut the height to 1,020mm.

Now to calculate the width, which requires us to work out what the circumference is. A quick calculation of   where D = diameter tells us that C, the circumference, of the finished column and therefore the minimum width of the cladding needs to be 1,100mm (rounding up):

3.142 × 350
= 1,099.7
Round up to 1,100mm

I'll add 20mm to this dimension to give us a bit of leeway when cladding. For more information on using formulae, refer to the 'Useful Geometry and Maths' section in Chapter 2. The cladding sheet should therefore be cut to 1,020mm height by 1,120mm width. The grain will need to run vertically (the height of the column) so that it can wrap around the formers.

Take the sub-frame and mark a centre line down one of the battens. This represents the starting point for the cladding, leaving us just enough material to fix the end to.

STEP 9

The key to success when cladding a curved surface like this is to break it down into sections. Begin by gluing up the first of the eight sections of the frame. Now carefully attach the starting edge of the cladding along the centre line on the batten, leaving an equal overhang on both the top and bottom edges, using a narrow stapler at approximately 40mm intervals. The overhang is important, as it is extremely difficult to clad a cylinder like this keeping the top and bottom edges flush all the time.

With the edge fixed down, work along the formers with the stapler until you reach the next batten, trying to work evenly as you go to keep

FLEXIBLE CLADDING

It is important to order the correct type of flexible plywood or MDF for curved cladding, depending on the dimensions of the piece you are constructing. In order for it to flex, the plywood will have most of its layers (plys) set with the grain running in the same direction. On an 8 by 4ft (2,440 × 1,220mm) sheet this can be supplied either 'long grain', where the sheet could be rolled to form an 8ft tube, or 'short grain' where it could be rolled to form a 4ft tube. The same applies to flexible MDF, which, instead of using multiple layers, relies on having one side of the sheet heavily 'kerfed' in one direction to allow it to bend – so double check your stock before you start!

Short grain makes a 4ft tube, long grain an 8ft tube – be sure to order the correct orientation for your project.

Step 9a: Double check the length of the cladding before going too far with the fixings.

the cladding from twisting, then staple down the cladding along the length of the batten. With the first section secured, double check the fit of the cladding by wrapping it around the formers as tightly as possible. If there is excessive overhang at the seam where the ends meet, this can now be trimmed to within 5–10mm of the desired length (depending on how brave you are) with a circular saw, by unwrapping the cladding as far as possible and clamping it down to the bench.

Continue working your way around the column, gluing and fixing one section at a time until you reach the final section.

STEP 10

Before applying any glue to the last section, push the cladding over to where it meets the starting edge to check the fit. If needs be, trim the edge back, either with a plane or Stanley knife and straight-edge if the cladding is thin enough, until the ends meet perfectly.

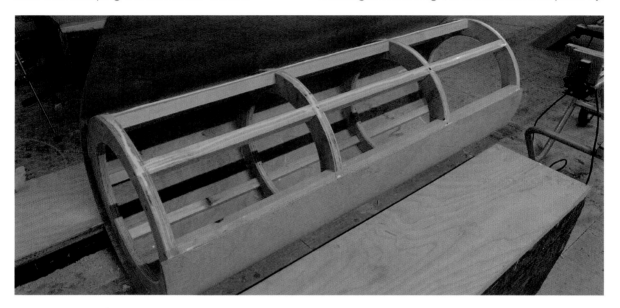

Step 9b: Work around the frame evenly, gluing up one section at a time.

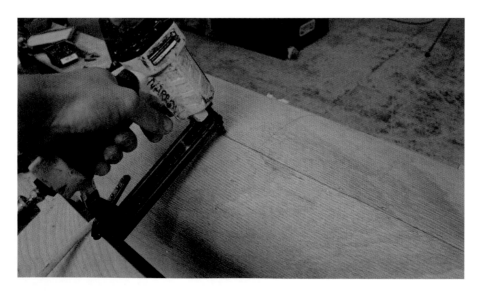

Step 10a: Fixing the seam.

Step 10b: Trim off the excess ply with a saw, or sand it down if there is only a fraction to remove.

The finished column.

Lift up the cladding, apply the glue, and then fix it down ensuring a neat join along the seam.

Any excess around the top and bottom can also be trimmed back, either by sanding it down or carefully cutting it flush with a saw. A Japanese-style pull saw is perfect for this kind of task as it allows a very flush cut to be achieved (and it looks really cool).

The seam and staple holes can now be filled and sanded back, and the column is complete.

GLOSSARY

ABTT The Association of British Theatre Technicians: 'We campaign on behalf of the theatre industry to ensure legislation is appropriate to the industry's needs, and that regulations are suitably drafted and enforced' http://www.abtt.org.uk (*see also* '*Yellow Book*').

arris To bevel the 90-degree edge on a piece of scenery with a block plane or router as part of the finishing process.

assembly or sub assembly Group of components already connected to aid further construction.

assembly area Large open area of a workshop or theatre specifically for assembling the set.

auditorium Area of a theatre where the audience sits.

axonometric Method of presenting three axes of orthographic projection in two dimensions without the use of perspective.

back set Distance from the face of a lock or latch to the centre of the spindle.

ball catch Simple fitting used to keep doors and windows closed with a sprung ball.

bench drawing or construction drawing Technical drawing specifically laid out and dimensioned to be built from. This will include material types, technical information relating to fixings and joints, and may include section and detail views.

bevel Angled edge or adjustable tool for marking angles (*see also* sliding bevel).

bomber hinge Adjustable sprung hinge (mind your fingers!) allowing a door to open in either direction and return to the centre of the opening.

boss plate Steel plate with a captive nut welded in the centre to allow scenery to be bolted together or to the stage.

brad Thin wire nail fired from a pneumatic gun or 'brad nailer'. Best suited to temporarily holding parts during assembly or discreetly fitting moulding.

breaking down Process of making a set or prop appear older or more worn than it actually is.

break-out Where timber splinters across the grain when it is machined or sawn.

butt joint Simple timber joint where two pieces butt together but do not interlock; usually fixed with screws, nails or wiggle pins; can require extra support.

castor wheel Used on stage trucks and scenery to aid movement, generally used in sets of four or more.

CDM Construction (Design and Management) Regulations 2015: UK legislation integrating health and safety into the design and construction planning and management process. More information can be found here: http://www.hse.gov.uk/entertainment/cdm-2015/introduction.htm

centre line Line on a ground plan running up/down stage indicating the centre; this is one of the most important datum lines when working on stage (see also setting line).

chamfer To put a bevel on an edge, or a small bevelled edge detail such as a 'stop chamfer'.

cladding The hard surface applied to flats and other scenic elements to cover large areas of framework and make them appear solid.

clenched nail A traditional method of fixing scarfe joints by using long nails that are then bent back on themselves to hold the joint securely.

CNC The general term for 'computer numerical control', and CAM or 'computer-aided manufacturing'; often used in scenery building in conjunction with a three-axis router or milling machine to mass produce components such as formers or complex fretwork.

convex bevel Method of sharpening chisels and planes involving honing a curved radius on to the end of tool.

corner plate A plate, often triangular, usually made of plywood used to reinforce butt joints.

corner block A plate, often triangular, usually made of plywood used to locate boards into position such as rostra tops on to frames.

cornice Horizontal decorative moulding at the junction of a wall and ceiling.

COSHH 'Control of Substances Hazardous to Health'.

countersink The tool for, or process of, drilling a bevelled hole for a screw head to sit into.

crosscut To cut across the grain of the wood as opposed to ripping along the grain.

cutting list Itemized list of the component parts taken from a bench drawing for the purpose of cutting to size.

development Drawing or mark-out of a surface unfolded, unrolled or flattened to find its true dimensions.

dimensions Numerical figures added to technical drawings indicating the size of components, usually expressed in mm (metric), or feet and inches (imperial) in the USA.

'dog and biscuit' A quick and simple way of building scenic flats using wiggle pins (dogs) and plates (biscuits).

door furniture The functional hardware added to a door such as handles, locks or numbers.

doorstop The thin timber strip applied around the door lining to prevent light leak and to stop the door from swinging through the frame past the closed position.

draughting (drafting, US) The process of producing technical drawings.

draughtsperson Person who specializes in producing technical drawings, either using a CAD or 3D design software package, or by physically drawing with pen and paper.

dressing set Items on set that are not used directly by the actors but serve to add detail.

durometer Scale of measurement determining the hardness of a material, commonly castor wheels in relation to scenery.

end grain The grain across the end of a piece of timber.

face side, edge The cleanest side or edge of timber components, usually chosen and marked to aid assembly and machining.

factory edge The uncut edge of a board, as it arrives from the supplier.

fit-up The process of building the set on stage ready for use by the production team.

fixings Hardware components such as screws and nuts and bolts, used to join materials together.

flats The 'walls' of the set, typically timber or steel frames clad with a thin sheet material to make them appear solid.

flush Two meeting surfaces being perfectly flat.

fly bar Part of a counterweight flying system – the tubular bar that can be raised or lowered during performance, and rigged with scenery and lights.

footprint Plan view of an object as it appears at stage or floor level. Used as a 1:1 mark-out for construction purposes.

formers Components of a three-dimensional object which have been cut to give it its shape; typically made from timber or plywood, and found in objects such as columns, ramps and cornices.

French brace Style of support frame for scenery, typically attached to the back of flats using backflap or pin hinges.

frieze Horizontal decorative strip around walls, usually flat or with minimal relief.

get-in The process of moving the scenic elements into the theatre and on to the stage.

'get-offs' Nickname for a set of treads leading off the stage.

'go off' Slang term for the process of adhesive curing or drying.

going Refers to the horizontal distance between the face of the first and last risers on a staircase, or on an individual tread.

grid The structural frame at the top of the fly tower, under which the fly bars sit when they are fully out.
ground plan The architectural floor plan of a theatre or performance space, usually drawn at stage level.

grub screw A small threaded bolt used for discretely securing hardware such as a door handle or bearing to a spindle or axle. It has an internal fitting for tightening with either a hex key or flat blade screwdriver.

half lap A timber joint where two pieces cross with an equal amount of material removed from each half to keep both faces flush.

haunch A small step machined into mortise and tenons, usually on the outside corners of a flat, to prevent the joint from slipping apart.
head and sill The top and bottom rail of a timber frame.

horn The waste material deliberately added to the end of the head and sill to protect the mortise during machining and assembly.

infill A bespoke section of rostra or deck, usually constructed to marry up the set with the edges of the stage.

isometric A method of presenting three axes of an orthographic projection in two dimensions using perspective.

jamb The internal face of a door frame.

jig An accessory to machines and tools designed to simplify the fabrication of multiple pieces.

kerf A cut, or series of cuts made to make removing or shaping material easier.

knot An imperfection found in timber where a branch had grown.

knuckle The tubular barrel on a hinge.

lapping The final stage of sharpening: the removal of the burr and final polish.

leaf The flat section of a hinge that is drilled and countersunk so screws will fit.

let in To remove material allowing hardware such as hinges to sit flush when fitted.

LEV local exhaust ventilation: an extraction system fitted to power tools and machinery for the purpose of reducing the levels of harmful dust and fumes from the workshop environment.

lock block A solid wooden block built into a hollow door behind the lock stile to allow door furniture to be properly secured.

lug A 90-degree steel bracket used to secure scenery together, typically welded or bolted on to framework.

machining The process of cutting or shaping components using workshop machinery or power tools prior to assembly.

mark-out The process of measuring and marking the positions of joints and adjoining components, or the position of the set on stage.

masking The black flats and softs used to hide the off-stage areas and technical equipment from the audience.

micro bevel A method of sharpening where a chisel or iron has a primary angle ground, and then a steeper secondary angle which becomes the sharp edge.

mirrored Elements that need to be constructed as a mirror image of each other or as a 'handed' pair.

model box The scale model produced by a designer of the finished set, usually 1:25 scale or occasionally 1:50 for larger venues; it will include furniture, paint finish and any scene changes that need to occur during the performance.

moulding decorative timber components such as an architrave or a dado rail.

noggin Slang for a small spacer in a frame that is too small to warrant being called a rail.

offcut The unused, or unusable waste material left over from cutting.

on edge A method of frame building where the narrow edge of the timber is set to the face to allow butt joints to be used.

orthographic projection A method of presenting technical drawings showing the top, front and side views.

oversail Where a material deliberately projects over an edge as opposed to being set flush.

PAR An acronym applied to timber indicating that it has been 'planed all round'.

paring The process of removing thin slivers of wood with a chisel.

peg A tapered wooden fixing traditionally used for securing mortise and tenon joints.

pilot hole A hole drilled slightly smaller than the diameter of the intended fixing, to help alignment and prevent splitting.

pin hinge A backflap-style hinge with a removable pin, useful for quick turnarounds.

plan view The 'top down' view on an isometric projection, usually drawn at base level.

Podgalug® (is a registered trademark of Flints) A lug "Designed and patented for use with the ratchet podger. Podgalugs are drilled with two holes, one for the podger to align the lug and the other for the bolt."

podger A ratchet spanner for tightening bolts and scaffolding clamps, with a tapered bar at one end to allow for easy alignment of bolt holes.

PPE personal protective equipment.

production manager The person overseeing and coordinating all technical aspects of a production, including the schedule, budget, and health and safety considerations.

production meeting A meeting for all heads of departments and the creative teams to discuss the progression of a production.

proud Misalignment, where a surface sits above its intended location.

push stick A tool used to hold or feed materials close to the blade on machinery.

rail The horizontal parts of a frame.

rake An angled stage, generally lower at the front and higher at the back, hence the terms 'up stage' and 'down stage'.

relief cut A cut made into the waste part of a piece to make further cuts easier.

rep, repertoire A theatre that shows a cyclical programme of productions, often changing every night. The Royal Opera House and the National Theatre are examples of companies that use the repertory system.

revolve A circular rotating stage platform used to move scenery or cast for different effects.

rip To cut timber along the grain; also a length of timber or board that has been cut in this way.

rise The measurement of the height of a step or staircase.

riser The vertical face of a step.

rostra A temporary platform used on stage.

run The timescale of a production's performances from opening night until it closes.

sacrificial fence A replaceable piece of timber fixed to workshop machinery to improve stability or reduce break-out.

saddle and block A pair of timber bearers for supporting a door on its edge; the saddle has a notch and wedge for securing the door in place while it is worked on.

sample Prototype of a method or finish for discussion prior to fabrication.

sash cramp A long adjustable metal clamp used in joinery.

scarfe joint A traditional method of joining long lengths of timber by creating a long tapered splice fixed with glue and nails.

scene change Where the scenery and props are changed between scenes during a performance to suggest a different location or time.

serge A wool-based felt, usually black for theatre use, used to cover masking flats and for soft masking such as borders. Chosen for its matt appearance, which effectively absorbs light.

set (on a sawblade) The angle at which the teeth of a saw blade protrude from the side of the blade in an alternating pattern.

set dressing Items on set that are not used directly by the actors but serve to add detail.

setting line The down-stage reference line, perpendicular to the centre line, used to determine the position of the set.

shoulder length The length of a tenoned rail excluding the tenons themselves – that is, the visible part of the rail when the frame is assembled.

shy Misalignment, where a surface sits below or just short of its intended location.

side elevation One of the views in an orthographic projection: a side view taken perpendicular to the plan and front elevation.

sightlines Marked on the ground plan and section to indicate the maximum view of the audience from the auditorium; used to plan masking and scenery placement via a sightline study.

size glue Also known as rabbit glue, used on canvas to help tension and seal it prior to painting.

spindle The square or threaded bar that fits through a door latch, on to which handles are fitted.

spigot A short tube designed to fit into a socket for the purpose of extending or connecting components or aligning scenery on stage.

square stage or English brace An extendable wooden brace useful for supporting smaller and lighter scenery; it features a hook at the top and a foot at the base for a stage weight.

star dowels Small star-shaped nails used to secure mortise and tenon joints.

stiffener A length of timber or steel added to the back of a flat or set in order to reduce torsional twisting or bowing and to keep surfaces flat.

stile The vertical parts of a timber frame such as a flat, door or window.

striking plate The metal faceplate fitted to the door frame into which the latch fits when closed.

string or stringer The former on a set of treads or staircase, which either takes the shape of the treads and risers (open string) or houses them (closed string).

strike To take the set down or remove scenery and props off stage during a scene change.

tenon The male half of a mortise and tenon joint, essentially a tongue cut in the centre of the end of a rail.

theming The process of aesthetically tying the set together in a stylized way, often accentuating shapes and textures.

title block The box on a technical drawing containing information such as the name of the project and designer, the scale, and any notes relating to materials or finish.

toggle shoe A tapered timber block, mortised in the centre to allow tenoned rails to be screwed into the stiles of a flat.

toolbox talk A team briefing at the start of each work day outlining the schedule and any additional health and safety considerations. An important part of the CDM process for set builders to follow.
trammel heads Adjustable compass heads that can be attached to a batten for marking out curves on a large scale.

trammel An arm for making an arc such as when using a router to cut a circle.

trap A door set into the floor of the stage to allow cast and crew access during a performance.
treads Sets of steps often used off stage by cast and crew. Single steps in a staircase are also referred to as treads.

truck A large base fitted with castors, on to which scenery can be fixed and moved as part of a performance.

tubular latch A cylindrical mortise latch fitted to internal doors, allowing the door to be opened by turning the handle.

void A hole cut into a former or flat for the purpose of saving weight or improving access.

weathering The process of making the set appear as though it has worn or corroded naturally over time.

white card The prototype model produced by a designer in the early stages of a production, mainly concerning the proportions and function of the design rather than the finish.

wiggle pin A corrugated fastener, also known as a 'dog', used to quickly attach two pieces of timber together quickly (see also dog and biscuit).

wing The sides of the stage area, often used for storing scenery and props for scene changes.

'Yellow Book' The nickname for ABTT's *Technical Standards for Places of Entertainment*.

FURTHER READING

Blaikie, Tim and Troubridge, Emma *Scenic Art and Construction: A Practical Guide* (The Crowood Press 2002)

Blurton, John Scenery: *Draughting and Construction* (Routledge 2001)

Fraser, Neil *A Backstage Bible* (Nick Hern Books 2018)

Holden, Alys, Samler, Bronislaw, Powers, Bradley L. and Schmidt, Steven A. *Structural Design for the Stage* (Focal Press 2015)

Thorne, Gary *Technical Drawing for Stage Design* (The Crowood Press 2010)

Woodworker's Journal, *Jigs and Fixtures for the Table Saw and Router* (Fox Chapel Publishing 2007)

USEFUL CONTACTS AND SUPPLIERS

ORGANIZATIONS

ABTT: The Association of British Theatre Technicians – abtt.org.uk

BECTU: The UK's media and entertainment trade union – bectu.org.uk

HSE: Information about health and safety at work and CDM – hse.gov.uk

National Theatre: nationaltheatre.org.uk

RADA: The Royal Academy of Dramatic Art – Acting and Technical Theatre training – rada.ac.uk

The Royal Opera House: roh.org.uk

SUSTAINABILITY

FSC: Forest Stewardship Council – fsc-uk.org

Julie's Bicycle: juliesbicycle.com – Improving sustainability across the creative sector

PEFC: Programme for the Endorsement of Forest Certification schemes

Scenery Salvage: scenerysalvage.com – Scenery recycling, hire and sales

Set Exchange: set-exchange.co.uk – A free message board for sharing sets and props

SUPPLIERS

Amari Plastics: amariplastics.com – Supply a wide range of polycarbonate as well as wide range of other plastic-based sheet materials

Creffields Flameproofed Boards: creffields.co.uk – Supply fire-retardant sheet materials for use on stage

Flints Hire and Supply: flints.co.uk – The single source for specialist theatre hardware and tools

J. D. McDougalls: mcdougall.co.uk – Suppliers of flame-retardant fabric such as canvas, serge and underfelt

Peter Evans Studios: peterevansstudios.co.uk – The main supplier for vacuum-formed plastics

Volund Timber Ltd: Great London-based supplier for timber, used by many scenery shops

INDEX

access 13, 20, 24, 28, 140, 181, 192, 197

accessible 134,186

angle 19, 28–9, 45, 47–9, 54, 71, 83, 88–94, 102, 114, 120, 122–6, 136, 143, 190–98

arc length 90

architectural 24, 201–15

area 20, 28, 60, 69, 75, 78, 83–4, 89, 95, 159

arris, arrising 58, 61, 106, 112, 114, 127, 136-138, 145, 175–6, 183, 195

audience 10, 14, 23, 25, 60, 63–4, 123, 148, 154, 167, 178

axonometric 19

backflap hinges 28,

bandsaw 42, 51-52, 66–7, 160, 208

bench 18, 21, 31, 35–6, 47, 54–5, 70, 75, 81, 83, 92, 103, 105, 112–14, 120, 123–7, 137, 142, 145, 156, 158-160, 175, 187, 197, 210, 214

bisect 92–3

board 16, 19, 22, 33, 51, 56, 58, 62–4, 78, 89, 116, 125, 132, 163, 181, 184, 191, 195, 197–9, 200–01, 205, 209–10

boss plate 30

brace 28, 60, 62, 94–5, 109, 114, 122–7, 148, 155, 157, 197

budget 9, 13, 21, 25, 33, 199, 200–01

butt hinge 28, 158–9

butt joint 58, 66, 99, 101-103, 107, 124, 126, 148, 150, 183–4

CAD 16, 18–19, 83

canvas 99, 106, 119–22

carpenter 9–10, 14, 16, 21, 35, 38–9, 44, 54, 60–61, 66, 78, 95, 99, 102

casement 169, 172

castor 28, 181, 185–8, 197

CDM (Construction Design and Management) 13, 23, 32

centre line 20, 76, 95, 104, 111, 166, 201, 212

chisel 36–8, 45-50, 54, 66, 69, 73, 76, 110, 112, 137, 143, 145, 150–51, 162-8, 174, 176, 205

chord 89

circle 83–90, 207–10

circumference 83, 86-88, 207, 212

clad, cladding 83, 99, 100–01, 105–6, 116–18, 129, 148, 151, 154-7, 179, 184, 205, 207, 212–15

clenched 73

closed string 140–41

CNC 172, 189–90, 200, 208,

coach bolt 29, 186–8

compound mitre saw 43

cornice 24, 204–6

countersink 38, 58–9, 77, 105, 137-138, 145, 151, 157, 163, 166, 168, 184, 193

cross cut 41

cutting in 31, 168, 192

cutting list 54–8, 62, 102–3, 107–8, 124, 141, 148–9, 153, 173, 183, 191, 211

design 9, 12, 14–17, 19, 21–3, 30, 99–100, 123, 139, 147, 155, 165, 169, 172,178, 189, 201, 204–5, 208

designer 12, 14–17, 20–21, 24–5, 155–6, 167

developed surface 19

diameter 83, 86, 185, 189, 207, 209, 212,

director 14, 16–17, 25, 155,

dock door 28

door 24, 28, 32, 69, 147–68

draughting (drafting) 21, 92, 171

draughtsperson 9

drawings 9, 14, 16–21, 31, 55

durability 21, 23, 63–5, 101,123, 199

dust 35–7, 54, 65, 209–10

ear defenders 35

elevation 18, 20, 181,

eye bolt 28

fit up 13–14, 20, 31, 53, 157–8, 201

flown 23–4, 79, 101, 157

flys, flying 20, 79, 101, 157, 205

folding gate 28, 189

footprint 20, 33, 58–9, 92, 94, 181, 183–4, 191–2, 197

foreman 13–14, 16

former 23, 29, 63, 83–4, 134, 179, 197, 204–14

get in 28, 31, 37, 56

get out 13–14, 32, 201

grain 62–4, 70, 111, 116–17, 124, 127, 137, 140, 149, 152–3, 155, 176, 199, 212–13
ground plan 14, 20, 31
grummet 79–81

haunch 107–12, 173
health and safety 13, 23, 35
hinges 28–30, 122, 155–64
hone 36, 45–8
horn 107, 111–15, 173
housing 141–5

imperial 22, 62
isometric 19

jigs 51–2, 59, 69–70, 163, 208, 210
joinery 9, 58, 66, 99, 147

label, labelling 30–31, 36, 54–6, 58, 61, 102–3, 107, 123, 149, 173, 183, 191
lap joint 73, 78, 148–9, 173
lapping 44–6, 49
line and cleat 29
line types 17–19
lock block 148–9, 151, 158
lug 29

machine screw 28, 30, 79, 81, 157
mark out spacer 109
masking 14, 99, 101, 119
MDF 22, 63-65, 78, 92, 116, 199, 213
method statement 32
micro bevel 44–9
model 14–17, 23
mortise-and-tenon 51–2, 56, 66, 73, 99-101, 106–7, 119, 152, 159, 172–3, 189
mortiser 43, 66
moulding 20, 24, 32–3, 64–5, 92, 148, 152–4, 175–6, 179, 203–6

on edge 56, 63, 78, 99, 101–3, 119, 126, 133, 135, 145, 148, 155, 159, 174, 205
open string 132, 139-140
orthographic 18–19, 21

paint 23, 25, 35–6, 40, 63, 106, 117–19, 127, 158, 160, 167, 199–201
PAR (Planed All Round) 62
peg 51–3, 66, 100, 112–15
pin hinge 28, 122, 164,
planer thicknesser 43
plotter 18
polycarbonate 65, 100, 110, 116, 153, 171–2, 176, 178, 199,
power tools 39, 44, 67, 190–91, 208
PPE Personal Protective Equipment 35-36, 54, 65
production manager 13–15, 21, 32
production meeting 16,
projection 9, 18–19, 48
props 13–14, 16,
Pythagoras' Theorum 83, 94–5, 124

rehearsal 27
riser 132–3, 137–40
rostra, rostrum 19, 22, 28–9, 51, 63, 188–97

saddle and block 160–61
sample 16, 205
sash 30, 112, 114, 150, 169–70, 172, 175
scale 9–10, 13, 14, 17–18, 20, 23, 33, 58, 123, 147, 195
scarfe 58, 69–73
scene change 14, 16, 29, 196
scenic art 13, 14, 21, 31, 117, 158, 199
schedule 13, 27
section 20, 31, 185
sector 84, 88-90
setting line 20, 201
sheet join 23, 78, 84, 151

shoulder 51, 77, 105, 108, 110, 112, 150, 151, 173–4
show floor 31, 199–201
sill 59, 107–11, 114, 123–6, 155–8, 179
sliding/adjustable bevel 38, 92, 126, 141–2
snagging 31
split batten 29
steel 24, 29, 33, 55, 79, 99, 101, 145, 157, 169, 189–90, 195
stiffener 101
stile 51, 55, 79–81, 94, 100, 103–5, 108–13, 123, 125, 129, 148-155, 159, 165, 173, 174
storage 27, 32, 69, 101, 127, 160, 186
string, stringer 23, 132–45, 198

T nailer 24
T nut 28, 79, 81, 157, 186
table saw 41, 48, 51, 69–70, 92, 135–6, 190–93
tenoner 43, 66–7, 109, 173
texture 20, 24–5, 65, 154, 158, 199
tie marks 31
title block 17,
toggle shoes 51, 107–9, 112–13
toolbox talk 14
touring 23
trammels 38, 84, 89, 92–3, 97, 134, 192, 207–9
transport 27–8, 31, 33, 69, 100
travelling batten 156–8
tubular latch 151, 164–5

underfelt 194–5

Vac-form 55, 65, 204, 206

wall saw 42
white card 14–15
wiggle pins 78, 100, 126